A GUESS AT THE RIDDLE

A GUESS AT THE RIDDLE

ESSAYS ON THE PHYSICAL UNDERPINNINGS
OF QUANTUM MECHANICS

David Z Albert

HARVARD UNIVERSITY PRESS
Cambridge, Massachusetts
London, England
2023

Copyright © 2023 by the President and Fellows of Harvard College
All rights reserved
Printed in the United States of America

First printing

Library of Congress Cataloging-in-Publication Data
Names: Albert, David Z, author.
Title: A guess at the riddle : essays on the physical underpinnings of
quantum mechanics / David Z Albert.
Description: Cambridge, Massachusetts ; London, England:
Harvard University Press, 2023. | Includes index.
Identifiers: LCCN 2022058369 | ISBN 9780674291263 (cloth)
Subjects: LCSH: Quantum theory. | Space. | Physical laws. | Reality. |
Wave functions. | Physics—Philosophy.
Classification: LCC QC174.125 .A42 2023 | DDC 530.12—
dc23/eng20230327
LC record available at https://lccn.loc.gov/2022058369

Oh Lana Turner we love you get up

CONTENTS

A Note to the Reader ix

Introduction *1*

1 A Guess at the Riddle *11*

2 Physical Laws and Physical Things *61*

3 The Still More Basic Question *86*

Acknowledgments 127

Index 129

A NOTE TO THE READER

One way of reading this book is as an epilogue—or (more precisely) as a denouement—to a book I wrote more than thirty years ago, which was called *Quantum Mechanics and Experience*. That earlier book was meant to do two things:

1) To introduce its readers to the elementary nuts and bolts of the foundations of quantum mechanics—the idea of superposition, the basics of the standard mathematical formalism of quantum mechanics, the argument of Einstein and Podolsky and Rosen to the effect that that formalism, considered as a description of the fundamental reality of the world, must be somehow incomplete, Bell's extension of that argument into a general and powerful theorem about locality, the elaboration (by figures like Erwin Schrödinger, and John von Neumann, and Eugene Wigner) of the quantum-mechanical measurement problem, and various proposals (theories of the spontaneous collapse of the wave-function, Bohmian mechanics, the so-called many-worlds interpretations of quantum mechanics, and so on) for solving it.
2) To show how the above-mentioned proposals for solving the measurement problem, taken together, amount to a

decisive rebuttal of the hugely influential arguments of Niels Bohr and his circle to the effect that the behaviors of subatomic particles—and (by extension) the inner workings of the world as a whole—were somehow not *susceptible* of any literal and objective and realistic and otherwise old-fashioned sort of scientific account.

But the question of how an account like that might actually *look,* the question of what an account like that might actually *say,* was left open at the end of that book—and *that,* as I will explain in more detail in the Introduction, will be the central topic of this new one.

There are parts of the present book (then) that are going to take it for granted that the reader has some familiarity with the elementary nuts and bolts that I listed above. And for those who do not—and it is to those readers, in particular, that this note is addressed—there are now, happily, any number of clear and concise and relatively painless introductions to consult. Aside from *Quantum Mechanics and Experience* itself, there are (for example) excellent books by Tim Maudlin, Jeff Barrett, and Travis Norsen.[1] There is also a wonderful fifty-page introduction to these matters by Alyssa Ney.[2] Any of these will leave the reader well prepared to follow the discussion here.

---

[1] Tim Maudlin, *Philosophy of Physics: Quantum Theory* (Princeton, NJ: Princeton University Press, 2019); Jeff Barrett, *The Conceptual Foundations of Quantum Mechanics* (Oxford: Oxford University Press, 2019); and Travis Norsen, *Foundations of Quantum Mechanics: An Exploration of the Physical Meaning of Quantum Theory* (Cham, Switzerland: Springer, 2017).

[2] Alyssa Ney, "Introduction," in *The Wave Function: Essays on the Metaphysics of Quantum Mechanics,* ed. Alyssa Ney and David Z Albert (Oxford: Oxford University Press, 2013), 1–51.

A GUESS AT THE RIDDLE

Introduction

Quantum mechanics, when it was first written down, was famously obscure about the circumstances in which one or another of the possible outcomes of a measurement actually *makes its appearance* in the world. All it said was that the outcome emerges "when the measurement is performed," and it offered no precise idea of what the phrase inside the quotation marks was supposed to mean.

You might have expected people to see that obscurity as a defect or an incompleteness in the quantum-mechanical formalism—you might have expected them to see it (that is) as something that needed to be *fixed*. But (as it actually happened) there was an enormously influential circle of physicists around Niels Bohr, at his institute in Copenhagen, who saw it very differently. It struck them as something revolutionary and profound. It seemed to them to point to a tension, the likes of which had never been encountered before, at the very heart of the scientific project itself.

The tension was supposed to run something like this: On the one hand, the language of classical physics was supposed to be indispensable to the business of *observing* and *recording* and *communicating* and *reasoning* about the *outcomes of experiments*—the language of classical physics was supposed to be indispensable (that is) to the general business of doing *empirical science*. And on the *other* hand, some of those experiments—the experiments (in particular) by means of which physicists had lately begun to explore the interior of the atom—were supposed to have shown that the language of classical physics was radically *unfit* for the business of actually *describing the world*.

And Bohr and his circle insisted that as soon as one had *appreciated* this predicament, as soon as one had taken the *full measure* of this predicament, it became clear that the obscurity I alluded to above was not *susceptible* of being fixed, and that the mechanism of the transition from the possible outcomes of a measurement to the actual outcome of that measurement was not going to *admit* of any detailed scientific account, and that the only way of making sense of the scientific project, and that the only way of going *on* with the scientific project, was to look at it through the lens of an especially militant version of instrumentalism. Science (on this way of thinking) is in no other or more expansive or more ambitious business than the business of *predicting* the outcomes of measurements. And the proper employment of words like "measurement" and "outcome" *themselves*, at least insofar as the discourse of empirical science is concerned, is as something like *primitive terms*. And so the business of analyzing how the elementary examples of measurement "work," the business of clarifying how it is that the outcomes of elementary measurements ever actually "make their appearance" in the world, if there is any such intelligible business at all, can certainly not be the business of empirical science. And so the goal of somehow "fixing" or "eliminating" the obscurity I alluded to above is simply *confused*, and the traditional aspiration of physics—the aspiration (that is) to offer us a true and objective and exhaustive and literal and realistic and mechanical account of *what the world is like*—will need to be *abandoned*.

And all of this quickly hardened into a rigid and powerful orthodoxy. As early as 1927 (for example) Werner Heisenberg and Max Born were prepared to declare that the standard quantum-mechanical formalism, with all of its accompanying obscurity, amounted to "a closed theory, whose fundamental physical and mathematical assumptions are no longer susceptible of any modification." And from then on, all sorts of questions about what things do, and how they work, were more or less universally declared to be nonsense, and the business of inquiring any further into these matters was actively and relentlessly and often brutally discouraged. Clear and simple and devastating critiques of the ideas of Bohr and his circle—from figures like Erwin Schrödinger and David Bohm and Hugh Everett and John Bell, and especially and particularly from Albert Einstein himself—were met with silence, or with derision, or with inexplicable misunderstanding, or answered with outright gibberish. Lives and careers were destroyed. And all of this persisted, in any number of different forms, and by any number of different means, and with undiminished zeal and intensity, for most of the previous century.

Bohr and his circle saw themselves as the vanguard of a brave and visionary and world-historical intellectual upheaval—and they saw figures like Einstein and Schrödinger and Everett and Bohm and Bell as somehow too *timid,* or too mired in what they called "classical ways of thinking," to keep up. But a case can be made that exactly the opposite was true. A case can be made (that is) that what had always been genuinely revolutionary in the scientific imagination was the original and unbounded and omnivorous and terrifying aspiration to reduce the entirety of the world to a vast concatenation of simple mechanical pushings and pullings. And what Bohr and his circle were up to (on this way of looking at things) was a profoundly *conservative* attempt to somehow *set a limit* to aspirations like that.

And there is something poignant and remarkable about all of this having happened at just the historical moment that it did. It was precisely in the 1920s, and precisely with the advent of quantum mechanics, that the possibility of a unified and comprehensive and thoroughly physical account of our everyday empirical experience of the world first began to take palpable shape. Such things had, of course, been dreamed of before. Boscovich and Laplace (for example) had been thinking, as far back as the eighteenth century, about what things might be like if the world as a whole were governed by the laws of something like Newtonian mechanics. But nobody had any idea, prior to the work of Faraday and Maxwell and Einstein, how phenomena like electricity and magnetism and light were going to fit into the picture. And nobody had any idea, prior to the discovery of quantum mechanics, how atoms worked, or *could* work, or what chemistry might be about. And it must have been impossible even to imagine, prior to the work of figures like Darwin and Wallace and Boltzmann, what a believable mechanical account of the origins and designs of living organisms was going to look like. And it was only sometime in the 1920s that one could begin to see, dimly (to be sure) but well enough, how relatively ordinary sorts of scientific ingenuity might someday actually clear all these impediments away. And it was then that the logical positivists began to think about assembling their *International Encyclopedia of Unified Science*. And it was then that H. P. Lovecraft remarked, "The sciences, each straining in its own direction, have hitherto harmed us little; but some day the piecing together of dissociated knowledge will open up such terrifying vistas of reality, and of our frightful position therein, that we shall either go mad from the revelation or flee from the light into the peace and safety of a new dark age." And you might say that it turns out to have been just that flight, and just that blindness, and just that dark age, that Bohr and his circle, at the last possible moment, at what apparently presented itself to many people as the very precipice of madness, bequeathed to us.[1]

Anyway, as everybody knows, everything that Bohr and his circle had to say about these matters turns out to have been wrong. And we are now aware of a number of very promising strategies for *modifying* or *completing* or *precisifying* the standard quantum-mechanical framework in such a way as to simply *eliminate* the obscurity about the process of measurement that I alluded to above. And with that obscurity out of the way, physics can again aspire to offer us an objective and literal and realistic and comprehensive and mechanical picture of *what the world is actually like*. And there has lately been a great deal of interest in the business of asking what quantum mechanics *itself* may have to *contribute* to a picture like that.

And that latter business is the business of this book.

⁓⁓⁓⁓⁓

Here are two ways one might go about looking into the question of what it is that quantum mechanics has to tell us about the fundamental structure of the world:

The "Top-Down" Approach

Start with the finished mathematical formalism of one or another of the strategies for eliminating the obscurity—start (that is) with the finished mathematical formalism of (say) Bohmian mechanics, or of Everettian mechanics, or of the Ghirardi-Rimini-Weber (GRW) theory, or what have you. And ask how that formalism can be *understood,* ask how that formalism can

---

[1] Adam Becker has provided a careful and readable and very engaging history of the debates alluded to over the past several paragraphs in *What Is Real? The Unfinished Quest for the Meaning of Quantum Mechanics* (New York: Basic Books, 2018).

be *interpreted*, as a literal, objective, comprehensive, mechanical, scientifically realist and empirically adequate representation of the world.

Consider (for example) the GRW theory. And ask: What is this theory *about*? What is to the GRW theory as *material particles* are to Newtonian mechanics? What is to the GRW theory as material particles and *electromagnetic fields* are to Maxwellian electrodynamics? What is it that the laws of time-evolution in the GRW theory are the laws of the time-evolution *of*? And the obvious answer, if you put the question that way, is *the quantum-mechanical wave-function*. And this suggests that the wave-function is the fundamental concrete object of the GRW theory—just as particles are the fundamental concrete objects of Newtonian mechanics, and just as particles and electromagnetic fields are the fundamental concrete objects of Maxwellian electrodynamics. And that, in turn, suggests that the fundamental physical *space* of the world is the very high-dimensional space in which the wave-function undulates—and that the three-dimensional space of our everyday *experience* of the world, and the particles that make their way about in that space, must somehow be understood as *emerging* from the pattern of those undulations. And all of this has always struck me as an eminently natural and sensible and believable way of making sense of a theory like GRW—and I have devoted a good deal of work, over the years, to proposing and elaborating and defending it.

But it is certainly not uncontroversial. And there are other proposed understandings of the GRW theory in which the fundamental concrete items are continuous mass-densities, or "flashes," or what have you, located in a fundamental *three-dimensional* physical space—and in which the metaphysical status of the *wave-function* is not that of a concrete fundamental physical *object*, but something more like that of a *law*. And there are closely analogous debates between so-called wave-function realist approaches and so-called primitive ontology approaches to Bohmian mechanics and Everettian mechanics as well.

Anyway—*all* of the above are examples of what I am calling "top-down" approaches. All of them begin (that is) with some finished quantum-mechanical formalism—the GRW theory, Bohmian mechanics, what have you—and treat that formalism as a sort of *skeleton* or *scaffolding* on which to drape a metaphysical picture of the world. And this is the approach that has always been adopted, in the scientific and philosophical literature, as far as I know, until now.[2]

The "Bottom-Up" Approach

Think of it this way: In the early years of the twentieth century, the incredibly strange behaviors of subatomic particles—which experiments were just then beginning to uncover[3]—presented a riddle: What could *behave* that way? What *are* these "particles"? What is the world (really, fundamentally) made of? And the attempts of Bohr and his circle to come to grips with this riddle somehow *got off on the wrong foot,* and all of a sudden Hegel and Kierkegaard and Buddhism and modernist crises of representation and principles of complementarity and verificationist theories of meaning and alternative systems of logic and interactive dualism and the cold war and the House Un-American Activities Committee and God knows what else got dragged into the conversation, and the result was a gigantic and protracted intellectual disaster. And what needs doing now is precisely the old-fashioned

---

[2] A nice sampling of such approaches in this "top-down" tradition can be found in Alyssa Ney and David Z Albert, eds., *The Wave Function: Essays on the Metaphysics of Quantum Mechanics* (Oxford: Oxford University Press, 2013).

[3] I am thinking here (for example) of the double-slit experiments, and the electron diffraction experiments, and the photoelectric effect, and the measurements of the emission spectrum of hydrogen, and so on.

everyday unparadoxical honest scientific work that Bohr and his circle ought to have done for us in the first place. The thing to do (that is) is to *start over:* to clear one's head, and to try to imagine, from scratch, what kinds of simple mechanical to-ings and fro-ings of what kinds of concrete fundamental physical *stuff* might explain why those experiments come out the way they do. Not to dandle the riddle, and not to enshrine it—but to *solve* it.

On *this* approach, the question of "interpreting" the formalism is simply not going to *come up*—or not (at any rate) in anything like the way it does in the top-down approaches—because it will have been precisely *by way* of such an interpretation, it will have been precisely by way (that is) of some flatfooted and literal and mechanical picture of what the inner workings of the world might actually be *like,* that we will have *arrived* at the formalism, if all goes well, in the *first* place.[4]

And I will have things to say about both of those approaches in the three essays that follow.

~~~

The first essay is concerned with the business of working one's way up from the bottom. This will lead us back, in the end, to a conception of the quantum-mechanical wave-function as concrete, high-dimensional, fundamental physical stuff—but it will

---

[4] Here's another way to put it: On this approach, if all goes well, the business of interpreting this or that finished mathematical formalism of quantum mechanics is going to look more or less like the business of interpreting the finished mathematical formalism of *Newtonian* mechanics. On this approach, if all goes well, the sorts of difficulties one encounters in interpreting this or that finished mathematical formalism of quantum mechanics is going to be more or less *on a par* with the sorts of difficulties one encounters in interpreting the finished mathematical formalism of Newtonian mechanics. There is no *shortage* of these latter sorts of difficulties, of course—but everybody seems to agree that the *acuteness* of these latter sorts of difficulties is of an entirely different *order,* of an entirely *lesser* order, than those that have traditionally been thought to confront us in making sense of *quantum mechanics*.

take us there by a route that feels much more *illuminating*, and much more *explanatory*, to me, than anything that's been on offer in the literature so far.

Here (in a nutshell) is how that's going to work: The whole business hinges on a distinction between two different conceptions of physical space. One of these identifies physical space with the set of points at which a *material particle* might in principle be located—and the other (which I will argue is more fundamental) identifies physical space with a set of points that satisfies (among other conditions) the condition that a specification of the physical situation at every one of those points amounts to a complete specification of the physical situation of the world. In classical physical theories—in Newtonian mechanics, and in Maxwellian electrodynamics, and in the classical versions of the special and general theories of relativity—these two spaces invariably *coincide*. But it turns out to be easy to invent very simple hypothetical physical systems for which they *come apart*, and it turns out that as soon as they *do* come apart there are paradigmatically quantum-mechanical sorts of behavior more or less *everywhere you look*, and there turns out to be a relatively straight line from these very simple systems to the familiar mathematical formalism of quantum mechanics itself.

The second essay is concerned with the debate between the "wave-function realists" and the "primitive ontologists"—which is to say that it is concerned with the business of working one's way *down* from the *top*. As I mentioned above, some of the strategies for *resisting* the thought that quantum-mechanical wave-functions are concrete high-dimensional fundamental physical stuff involve treating the wave-function as something *nomic*. And this immediately presents us with what (I guess) is the central question of the interpretation of *any* fundamental physical theory—which runs something like this: Given the finished mathematical formalism of some fundamental physical theory, how do we *tell*, and how *can* we tell, and how *should* we tell, which elements of that formalism are to be treated as representing concrete physical *stuff*, and which are to be treated as representing something *else*—something (maybe) more like a *law*?[5] Questions like that have, of course, been around from the earliest beginnings of the modern physical project—in the debate between Leibnitz and Clarke (for example) about the metaphysical status of *motion* and *space*, and in nineteenth-century discussions about the metaphysical status of *fields*, and so on—but they come up in a new and more general and more radical and more urgent way in connection with the quantum-mechanical wave-function. And that's the question that this second essay—"Physical Laws and Physical Things"—is about.

And the third essay is taken up with a painful and ungainly struggle with a still more basic question—a question that is (you might say) *prior to* and *deeper than* all of the questions that are considered in the earlier two essays—the question (in particular) of whether or not it is even susceptible of being coherently *entertained*, the question (that is) of whether or not it even amounts to a conceptual or metaphysical *possibility*, that the to-ings and fro-ings of some fundamental high-dimensional wave-function-stuff give rise to the tables and chairs and rocks and trees and haircuts and universities of our everyday experience of the world. The first essay, and the second essay, and the whole project of so-called wave-function realism, all take it for granted that the answer to that question is *yes*. But there have always been people whose guts told them otherwise. And there is (it seems to me) a real and profound and difficult question there.

---

[5] Of course, there are also questions about which elements of the formalism are not to be treated as representational *at all*—coordinate systems, for example, and gauge degrees of freedom, and so on. Those questions are not going to concern us here—or (rather) they are going to concern us only insofar as they turn out to have an impact on this question of the distinction between physical *things* and physical *laws*.

1

A Guess at the Riddle

The thought here, the "guess" of the title, is that everything that has always struck everybody as *weird* about quantum mechanics—the *essence* (you might say) of the quantum-mechanical, or maybe the essence of the difference between the quantum mechanical and the classical—has to do with a certain previously unappreciated ambiguity in our conception of space. That's what this essay is meant to explain, and to make plausible.

Constructions with a Pair of Particles

Let me begin by introducing a useful mathematical device with which some readers might be unfamiliar.

Every classical physical system can be uniquely associated with a formula, called its Hamiltonian, that expresses the total *energy* of the system in question—that expresses (that is) the sum of the kinetic and the potential parts of the energy of the system in question—as a function of the values of its physical degrees of freedom.

And it turns out, and this is why the Hamiltonian is such a useful device, that the Hamiltonian of a classical system concisely encodes everything there is to say about the dynamical laws of motion that that system obeys. It turns out (that is) that the way that the total energy of such a system depends on its degrees of freedom uniquely determines the equations of the evolutions of the values of those degrees of freedom in time. It turns out (that is) that there is a direct and straightforward and fully algorithmic procedure for deriving those equations—for any classical system—from its Hamiltonian.

To say that the Hamiltonian tells us everything about the *dynamical laws* of a classical system (however) is not quite to say that it tells us exactly what *kind* of a classical system it is that we are dealing with. Consider, for example, a very simple Hamiltonian—one that consists exclusively of kinetic energy terms—like:[1]

$$H = \tfrac{1}{2}m(dx_1(t)/dt)^2 + \tfrac{1}{2}m(dx_2(t)/dt)^2 \qquad (1)$$

This Hamiltonian fixes the dynamical laws of a system with two degrees of freedom—it fixes (that is) the differential equations that the two $x_i(t)$ need, as a matter of fundamental physical law, to satisfy. But there are two quite different sorts of physical system that a Hamiltonian like this one could very naturally be read as describing. We could read it (that is) as describing a *pair* of particles, both of mass m, moving around, in the absence of any forces, and without interacting in any way with each other, in a *one*-dimensional space. Or we could read it as describing a *single* particle, of mass m, moving around, in the absence of any

---

[1] Students of analytical mechanics will immediately recognize that the so-called Hamiltonian in equation (1) is not really a *proper* Hamiltonian *at all*. Hamiltonians are functions of x_i and p_i—considered as independent dynamical degrees of freedom—and emphatically *not* of quantities like dx_i/dt. The expression in (1) is actually the result of *starting* with a proper Hamiltonian, and then invoking the principle of least action, and then solving for the p_i in terms of the dx_i/dt, and then substituting the results into the original expression for H in terms of x_i and p_i. But expressions like (1), notwithstanding their various mathematical improprieties, turn out to provide a concise and convenient and perspicuous way of *picking out* a perfectly definite set of *dynamical equations of motion*—and that's all we're going to want from them here, or elsewhere in this essay, or anywhere in this book.

forces, in an otherwise empty *two*-dimensional space. All that the *Hamiltonian* does is to determine the differential equations that each of the two $x_i(t)$ need to satisfy. All that the Hamiltonian does—in this particular case—is to determine that

$$x_i(t) = a_i + v_i t \qquad (2)$$

where a_i and v_i are can be any real numbers. And that's precisely the sort of behavior that we would intuitively expect of *either one* of the two different physical systems described above.

~~~~~

Good. Let's make things a little more complicated.

Consider (again) a classical universe with two physical degrees of freedom—x_1 and x_2—but now suppose that the values of x_1 and x_2 evolve in time in accord with the Hamiltonian

$$\begin{aligned}H =\ & (1/2)m_1(dx_1(t)/dt)^2 \\ & + (1/2)m_2(dx_2(t)/dt)^2 + \delta(x_1 - x_2).\end{aligned} \qquad (3)$$

This Hamiltonian differs from the one in equation (1) in two important ways: it includes a very simple potential-energy term—$\delta(x_1 - x_2)$—and it allows for the possibility that the values of m in the two *kinetic*-energy terms might be *different*.

Here are two different ways of describing a universe whose dynamical laws are given by a Hamiltonian like the one in equation (3):

A) The universe consists of two point-like physical items, moving around in a one-dimensional space, and interacting with each other, on contact, by means of elastic collisions.

B) The universe consists of a *single* point-like physical item, moving around in a *two*-dimensional space, with an infinite potential barrier along its $x_1 = x_2$ diagonal—as in Figure 1.1.

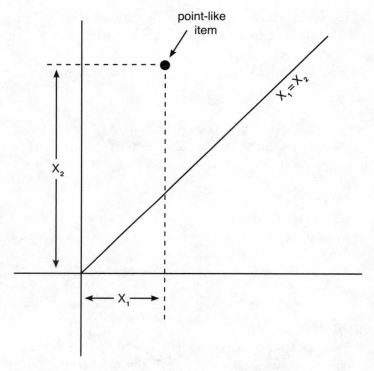

FIGURE I.I

These two descriptions—like the two descriptions we considered in connection with the Hamiltonian in equation (1)—are fully mathematically isomorphic to each other. But in this case, unlike in the previous one, the two descriptions are not apt to strike us as equally *natural*. Take almost anybody, with almost any kind of an education in physics, and wake them up in the middle of the night, and ask them to describe the sort of world that might have a Hamiltonian like the one in equation (3) as its fundamental law of motion—and you are likely to get something that's much closer to the language of description (A) than it is to the language of description (B). And the reasons for that will be worth pausing over, and thinking about.

To begin with, the mass associated with the kinetic energy of the x_1-motion and the mass associated with the kinetic energy of the x_2-motion, in the example we are considering here, are *different*. And we are used to associating a *single* mass with a single material object. You might even say that it is part and parcel of our very *idea* of what it *is* to be an "ordinary material object" that every such object is invariably associated with some single, determinate, value of its mass. And our everyday conception of *space* seems to have something to do with its being the *habitation* of objects like that. Our everyday conception of space (that is) seems to have something to do with the set of points at which an ordinary material object might in principle be *located*, or with the *stage* on which such objects seem to make their way about.

Good. But what if the masses happen to be the same? Won't it be just as natural (in *that* case) to think of this universe as consisting of a single material particle, moving around in a two-dimensional space, with an infinite potential barrier along the diagonal line $x_1 = x_2$?

Well, no. There are other issues here as well. It seems to be an important part of our everyday conception of the space in which material particles make their way about (for example) that it is both homogeneous and isotropic.[2] It seems to be an important part of our everyday conception of the space in which material particles make their way about (that is) that it should be

---

[2] This could obviously do with some qualification. Aristotle (for example) famously thought otherwise. But there is an intuitive and well-known and long-standing classical-mechanical conception of space that I am gesturing at here, which, I take it, is recognizable to everyone, and which (with a little work) can be formulated in such a way as to apply to special and general relativity as well.

just as easy, insofar as the fundamental laws of physics are concerned, for a material particle to be in one location as it is for it to be in another, and that it should be just as easy, insofar as the fundamental laws of physics are concerned, for a material particle to be moving in one *direction* as it is for it to be moving in another. And the *two*-dimensional picture of the sort of world we are considering here obviously features a fundamental law that distinguishes between points *on* the diagonal and points *off* of it. But if you look at that same law in the context of the *one-dimensional* picture—if you look (that is) at the potential term in the Hamiltonian in the context of the one-dimensional picture—all it says is that the two particles can't *pass through* one another. And *that* way of putting it obviously makes no distinction whatever between any two points in the one-dimensional space, or among either of its two directions.

Why couldn't we think of the presence of the potential barrier in the two-dimensional picture (then) not as a matter of fundamental law but (instead) as arising from the merely *de facto* configuration of a *field*? Well, *that* would amount to denying that the Hamiltonian in equation (3) is in fact the fundamental Hamiltonian of the universe. In that case (to put it in a slightly different way) the fundamental Hamiltonian of the world is going to be something more elaborate than the one in equation (3), something that offers a dynamical account not only of the evolutions of the coordinates x_1 and x_2, but also of the configurations of the *fields* (something, that is, that answers questions about how the fields *got* there, and how they *evolve*, how they are affected by changes in the x_1 and x_2 degrees of freedom, and so on). And that new fundamental theory is going to bring with it all sorts of new physical possibilities, and new counterfactual relations, that were not present in the original two-dimensional Hamiltonian that we were dealing with above.

So, what feels more familiar about the first of these descriptions is that it features a space that is homogeneous and isotropic, and that consists of the sorts of points at which ordinary material particles—particles (that is) that are associated with unique and determinate quantities of *mass*—might in principle be located. Let's refer to spaces like that (then) as spaces of *ordinary material bodies*. And we will refer to the material *inhabitants* of spaces like that as *particles*, or *material particles*, or *bodies*, or other familiar and flatfooted things like that—while reserving the more abstract and noncommittal term "item" for the inhabitants of spaces *other* than the space of ordinary material bodies. The two-dimensional space (on the other hand) is an abstract mathematical object that physicists refer to as the *configuration space* of a universe like the one we have been considering here—a space (that is) in which each of the points represents a possible geometrical *arrangement* of the particles in the familiar physical space of ordinary material bodies. Unlike the space of ordinary material bodies, the configuration space is merely an elegant device for bookkeeping, and not at all the sort of thing that anybody is ever tempted to take metaphysically seriously.

~~~~~

Good. Let's make things a little *more* complicated. Focus on the second of the two descriptions—the two-dimensional one—of the simple universe that we were talking about above. And now consider a *different* universe, a slightly more complicated one, that we obtain by introducing a *second* point-like physical item into the two-dimensional space.

This—while it is easy enough to *say*, while it is easy enough to *stipulate*—is not the sort of thing that we are accustomed to being asked to imagine in discussions of the mechanics of classical particles. In the previous example, the two-dimensional space was just an abstract way of keeping track of the configuration of the pair of familiar material particles in the one-dimensional space, and the point-like item that *floats around* in the two-dimensional space represented the *actual* configuration, at the time in question, of those particles. But a given set particles can, of course, have only *one* actual configuration at any one time—so the introduction of a *second* point-like item into the two-dimensional space is obviously going to turn that space itself into an altogether different animal, and we are going to need to feel our way around it all over again.

Let's take it one step at a time. We add a second point-like item to the two-dimensional space. And we stipulate that this second item is intrinsically identical to the first, and that it floats around in accord with exactly the same sort of Hamiltonian as the one in equation (3). The complete Hamiltonian of a universe like that (then) is going to be:

$$H = (1/2)m_1(dx_1(t)/dt)^2 + (1/2)m_2(dx_2(t)/dt)^2 \\ + (1/2)m_3(dx_3(t)/dt)^2 + (1/2)m_4(dx_4(t)/dt)^2 \\ + \delta(x_1 - x_2) + \delta(x_3 - x_4) \quad (4)$$

where $m_1 = m_3$ and $m_2 = m_4$. The situation in the two-dimensional space is depicted in Figure 1.2—and note that I have amended the notation a bit in order to facilitate the discussion to come: The two coordinate axes in the two-dimensional space are now denoted by the Greek letters λ and μ, and x_1 and x_2 are the λ and μ coordinates of item #1, and x_3 and x_4 are the λ and μ coordinates of item #2, respectively.

And if we proceed by analogy with the previous example, then we are going to be inclined to say that the universe depicted in Figure 1.2 can also be looked at as a *one*-dimensional space in which *four* particles—call them particle 1 and particle 2 and particle 3 and particle 4—are moving around. And note that

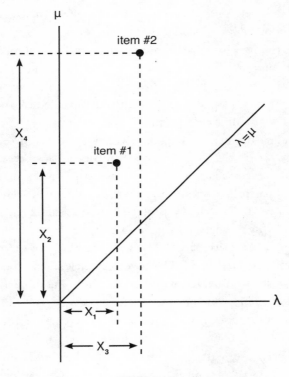

FIGURE I.2

the addition of this second point-like physical item to the two-dimensional space will have no effect at all on the *homogeneity* or the *isotropy* of the one-dimensional space in which those four particles move. There will, of course, be two more of those particles than there were before—but every one of the points in the one-dimensional space in which those four particles are moving around is going to have exactly the same physical properties as every *other* one of the points in that space, and the physics of moving in one direction in that space is going to be just the same as the physics of moving in the other direction. And so the one-dimensional space is still going to count here as a space of *ordinary material bodies*.

But the *way* in which those particles move around in that space turns out to be kind of funny. The way in which those particles move around in that space presents, on the face of it, a kind of paradox. Note (to begin with) that particles 1 and 3 are related in qualitatively identical ways—they are related (in particular) by means of qualitatively identical *mathematical projections*—to the two qualitatively identical *point-like items* in the *two*-dimensional space. And so it's going to be natural to think of particles 1 and 3 *themselves* as qualitatively identical to *each other*. And exactly the same goes for particles 2 and 4. And yet (and this is what looks paradoxical) the way in which particle 1 *interacts* with particle *2* is going to be *different* from the way in which particle *3* interacts with particle 2, and the way in which particle 2 interacts with particle 1 is going to be different from the way in which particle 4 interacts with particle 1. Particle 1, for example, is going to *bounce off* of particle 2, but it is also going to *pass right through* particle 4—and particle 4 is going to bounce off of particle 3, but it will pass right through particle *1*. And so, unlike in the two-particle case we considered before, a qualitative description of the physical situation of this world, at some particular time, in the one-dimensional space (that is: a complete specification of which four points in this one-dimensional space are occupied by particles, together with a specification of the *velocities* of the particles at each of those points, together with a specification of the *intrinsic properties* of the particles at each of those points) is *not* going to give us enough information to predict, even in principle, the qualitative situation of this world at other times.

The way out of the paradox is (of course) simply to look back at the *two*-dimensional version of this universe. When we picture this universe to ourselves as consisting of *two* point-like physical items floating around in a *two*-dimensional space, then everything immediately snaps back into place: a complete specification of the qualitative situation, at any particular instant, in the *two*-dimensional space (that is) *is* going to give us enough information to predict, in principle, how that situation is going to evolve into the future. And from *that* (of course) we are going to be able to read off all of the future qualitative situations in the *one*-dimensional space as well.

In the case we considered before, the one-dimensional representation of the universe and the two-dimensional representation of the universe were straightforwardly *isomorphic* to each other. In the case we considered *before* (that is) there was exactly *one* possible state of the point-like item floating around in the *two*-dimensional space corresponding to every individual one of the possible states of the two *material particles* floating around in the *one*-dimensional space. But *here* (as shown in Figure 1.3) there are *two* qualitatively different states of the two point-like physical items floating around in the two-dimensional space corresponding to every individual qualitative state the four material particles floating around in the *one*-dimensional space.

And so the *history* of the universe we are dealing with here—the history (that is) of this particular pair of point-like physical items floating around in this particular two-dimensional space—simply cannot be *presented* in the form of a history of the motions of ordinary material bodies, and the *dynamical laws* of a universe like the one we are dealing with here simply cannot be *written down* in the form of *laws* of the motions of ordinary material bodies. In the case we considered before (to put it slightly differently) the elementary physical *constituents* of the world—the concrete physical things on whose history the history of everything else *supervenes,* the concrete physical things to which the *fundamental dynamical* laws apply—were *ordinary material particles*. But the elementary physical constituents of a

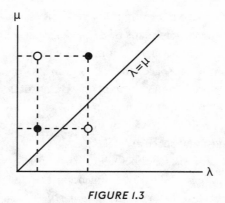

FIGURE I.3

Two qualitatively different configurations of the pair of point-like items in the two-dimensional space (one of which is represented by the two black dots, and the other by the two white ones) that are qualitatively identical in the one-dimensional space.

world like *this* one are obviously the *point-like items* in the *two-dimensional* space—and the reason everything looks so odd as viewed from the perspective of the *one-dimensional* space is that the *one*-dimensional space *isn't where things are really going on*, and the *material particles* that *move around* in that space are really just "shadows" (as it were) of the actual, concrete, fundamental physical items.

So there are apparently going to be *two* sorts of space that are worth taking seriously in a universe like this one. There is the *one*-dimensional space in which the *ordinary material bodies* make their way about, and there is the *two*-dimensional space in which the *elementary physical constituents of the world* make their way about. And the two-dimensional space is now the only one in which one can keep track of everything that's going on merely by saying what it is that's going on at every individual one of its *points*—it's the space (you might say) of *the totality of atomic opportunities for things, at any particular temporal instant, to be one way or another*. Let's call it "the space of the elementary physical determinables."

The space of ordinary material bodies and the space of elementary physical determinables turn out to be very different kinds of things. It is part and parcel of our idea of the space of ordinary material bodies (for example) that all of the points in it are going to be intrinsically identical to one another—but the above example makes it clear that we should have no such expectations, as a general matter, about the space of elementary physical determinables. The space of ordinary material bodies is the set of points at which you could imagine, in principle, placing the tip of your finger. But the items that make their way about in the space of the elementary physical determinables, at least in the case of the sort of universe we are considering now, are not material bodies at all.

But (notwithstanding all that) the space of the elementary physical determinables is clearly the more *fundamental* of the two. The situation in the space of ordinary material bodies (once again) supervenes, by definition, on the situation in the space of the elementary physical determinables—but the reverse is, of course, not true—or not (at any rate) in the sort of world we are thinking of here. So the sorts of distinctions that one can make in the language of the space of the elementary physical determinables are more *fine-grained* than the sorts of distinctions one can make in the language of the space of ordinary material bodies. Moreover, the space of the elementary physical determinables is what fixes the *elementary kinematical possibilities* of the world—and so it is (in that sense) something logically *prior* to the laws of dynamics, it's something like the *arena* within which those laws *act*. But the space of *ordinary material things* (as I mentioned before) is a space whose topology and whose geometry and whose very *existence* all explicitly *depend* on what the fundamental dynamical laws actually happen to *be*—it's something that the dynamics can be thought of as *producing*, something that is (in that sense) *emergent*.

The image of "space" that all of us grew up with (then) turns out to be a crude and undifferentiated amalgam of (on the one hand) the space in which the *ordinary material bodies* make their way about and (on the other) the space in which the *elementary physical constituents of the world* make their way about. That physics should never heretofore have taken note of the distinction *between* these two sorts of spaces is entirely unsurprising—because they happen to be *identical* (just as they were in the two-particle, one-dimensional example we considered above) in Newtonian mechanics, and in Maxwellian electrodynamics, and in the physics of everyday macroscopic practical life. The manifest image of the world (you might say) includes *both* a space of ordinary material things *and* a space of the elementary physical determinables—together with the stipulation that they are, in fact, *exactly the same thing*. And classical physics never gave us any reason to imagine otherwise. But (notwithstanding all that) these two ideas are worth carefully prying apart. They have nothing *logically* to do with one another, and it is the easiest thing in the world (as we have just seen) to imagine universes, and to write down Hamiltonians, in which (for example) they have different numbers of dimensions.

~~~~~

Ok. Let's get back, with all this in mind, to the particular system we were thinking about before—the one described by the Hamiltonian in equation (4). One of the effects of introducing a second point-like physical item into the two-dimensional space is (as we have seen) to *pry apart* the space of ordinary material bodies and the space of elementary physical determinables—to make them (in particular) into two distinct and topologically different spaces. And one of the effects of this coming-apart is that the goings-on in the space of ordinary material bodies—or (rather) the goings-on in the *physical universe,* as *viewed* from the *perspective* of the space of ordinary material bodies—look odd.

Particles 1 and 2 bounce off one another, and particles 3 and 4 bounce off one another, but (even though everything about the structure of the universe we are considering here suggests that particle 1 is intrinsically identical to particle 3, and that particle 2 is intrinsically identical to particle 4) particles 1 and 2 move around as if particles 3 and 4 simply did not exist, and particles 3 and 4 move around as if particles 1 and 2 simply did not exist. And so what we are presented with, in the space of the ordinary material bodies of a universe like this one, is maybe less like a collection of *four* particles floating around in a one-dimensional space than it is (say) like a pair of causally unconnected parallel *worlds,* in each of which there are *two* particles floating around in a one-dimensional space. Or maybe (again) it's like a pair of different *scenarios,* or like a pair of different *possibilities,* about how two *particular* particles—call them particle a and particle b—*might* be floating around in a one-dimensional space, or something like that. So we might say that there are old-fashioned fully determinate *facts* of the matter about those features of the history of the world on which the two possibilities *agree*—and that there somehow *fail* to be such facts about those features of the history of the world on which those two possibilities do *not* agree. We might say (for example) that when $x_1 = x_3$ and $x_2 \neq x_4$, then there is a determinate fact of the matter about the one-dimensional position of particle a (namely: $x_a = x_1 = x_3$), and there is *not* a determinate fact of the matter about the one-dimensional position of particle b.

If we were adamant about representing a universe like this to ourselves in its one-dimensional space of ordinary material bodies, we might do so with the help of an additional piece of notation—a pair of brackets (say), one of which links particle 1 with particle 2, and the other of which links particle 3 with particle 4—as in Figure 1.4—to indicate which particles share these "scenarios" with one another and which don't.

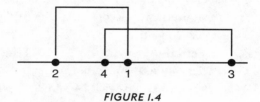

FIGURE 1.4

From the point of view of the *two*-dimensional space of *elementary physical determinables*, the brackets are just a way of keeping track of the connections between the four ordinary material particles in the one-dimensional space and the two point-like physical items in the two-dimensional space. But if we are resolute in banishing any thought of that latter space from our minds, then we are apparently going to need to think of the brackets as signifying some real and radically unfamiliar and not-further-analyzable physical connection between pairs of material particles *themselves*—something that can not be *reduced* to, something that does not *supervene* on, the spatial distribution of local physical properties.

All of this talk, it goes without saying, is figurative, and provisional, and defeasible. We are playing. We are feeling our way. And we will see, in good time, whether any of it has anything interesting to do with the world.

Let's make things more complicated again. And *this* move (by the way) will turn out to be the decisive one. Suppose that we were to add a term of the form $\delta(x_1-x_3)\delta(x_2-x_4)$ to the Hamiltonian in equation (4), so that it looks like this:

$$H = (1/2)m_1(dx_1(t)/dt)^2 \\
+ (1/2)m_2(dx_2(t)/dt)^2 + (1/2)m_3(dx_3(t)/dt)^2 \\
+ (1/2)m_4(dx_4(t)/dt)^2 + \delta(x_1-x_2) \\
+ \delta(x_3-x_4) + \delta(x_1-x_3)\delta(x_2-x_4) \qquad (5)$$

That would amount to adding a new and funny kind of an interaction—an interaction, not between two of the *particles* floating around in the *material* space, but (instead) between the two point-like *items* floating around in the *determinable* space—an interaction (that is) between what might previously have looked to us, from the perspective of the *material* space, like two different *possibilities,* or two different *scenarios,* or two distinct and parallel *worlds.*

Note (to begin with) that this new term is still going to preserve the invariance of the Hamiltonian under translations in the one-dimensional space—and so the material space of this new world, the space in which all points are intrinsically identical, the space in which particles have unique determinate masses, is still going to be one-dimensional. But the *behaviors* of these particles, as viewed from the one-dimensional space in which they live, are going to change, in interesting ways, yet again.

The effect of adding this new interaction is going to be quantitatively small—because collisions between the two point-like items in the determinable space are going to be much, much rarer than collisions between either one of them and the fixed diagonal potential barrier—but it is nonetheless going to be conceptually profound. From the perspective of the material space, things are still going to look more or less as if there are two pairs of particles floating around in two parallel possible situations—linked together by their mysterious brackets. But a more detailed examination is now going to reveal that this picture of parallel possible situations does not quite hold up—because the evolutions of these two possibilities can sometimes, in fact, *interfere* with each other.

Moreover, the effects of this new interaction, as viewed from the perspective of the space of ordinary material bodies, are going to be bizarrely *nonlocal*. Particles 1 and 3 are going to collide with one another (that is: particles 1 and 3 are going to *interact* with one another, particles 1 and 3 are suddenly going to become *visible* to one another, particles 1 and 3 are suddenly going to be unable to *pass through* one another) *only* in the event that particles 2 and 4 happen to be colliding with one another, somewhere in the material space, *anywhere* in the material space, at exactly the same temporal instant. And vice versa. And the mechanism whereby those two collisions make each other possible does not depend in any way whatever on the one-dimensional physical distance *between* them—it depends only on their primitive and unanalyzable and now even more mysterious bracket-connections.

And all of this should by now have begun to remind the reader of quantum mechanics. But it will be worth pausing for a minute over the question of *what* (in particular) is supposed to remind you of *what*.

There are a number of natural initial misunderstandings that need to be avoided: The introduction of an additional point-like concrete fundamental physical item into the two-dimensional space should not be confused (for example) with the introduction of an additional *particle* in the one-dimensional space, or with the introduction of an additional *pair* of particles in the one-dimensional space, or with the introduction of something along the lines of an additional "marvelous point" into a Bohmian-mechanical formulation of quantum mechanics.

It would be a little closer to the mark to say (as we were saying, a few pages back, about a slightly simpler imaginary universe than the one we are dealing with now) that the introduction of a second point-like item into the two-dimensional space is like the introduction of some additional one-dimensional *world*—some *parallel* or *possible* or *alternate* one-dimensional *world*—that the single original pair of particles somehow *also* (but *differently*) *inhabits*. But (as we have just seen) the fact that the two point-like items in the two-dimensional space can now *interact* with one another shows that thinking of them as parallel or possible or alternate one-dimensional worlds can't be quite right either.

The analogy that it will be helpful to have in mind, going forward, is (instead) this: Introducing another point-like item into the two-dimensional space is like adding another *term*, like adding another *branch*, to the quantum-mechanical *wavefunction* of the *single, original, two-particle system* in the one-dimensional space.

Consider (for example) a pair of spinless quantum-mechanical particles whose quantum state, at a certain time, is:

$$(1/\sqrt{2})[x=\alpha>_1[x=\gamma>_2 - (1/\sqrt{2})[x=\beta>_1[x=\delta>_2 \qquad (6)$$

If $\alpha \neq \beta$ (which is to say: if the two versions of particle 1 do not happen to be located at the same point in the three-dimensional space of ordinary material bodies) then the reduced density matrix of particle 2 will be an incoherent mixture of one state in which particle 2 is localized at γ and another state in which particle 2 is localized at δ—so if those two localized states are subsequently allowed to spread out, and to overlap with each other in three-dimensional space, they will not in any way *interfere* with each other. But if $\alpha = \beta = \zeta$ (which is to say: if the two versions of particle 1 *do* happen to be located at the same point—the point ζ—in familiar three-dimensional space) *then* particle 2 will be in the *pure* state $(1/\sqrt{2})[x=\gamma>_2 - (1/\sqrt{2})[x=\delta>_2$—so if *those* two localized states are subsequently allowed to spread out, and to overlap with each other in three-dimensional space, they will *very much* interfere with each other. And note (as well) that none of this depends in any way on the three-dimensional distance, or on the prevailing physical circumstances, *between* ζ (on the one hand) and γ and δ (on the other).

Let's follow that thought a little further.

You might say—or somebody might say—that so long as the two particle-*1* branches are kept apart, then the two particle-*2* branches are *invisible* to each other—and that as soon as the two particle-1 branches are brought together, then the two particle-2 branches are suddenly able to *see,* or *affect,* or *interfere* with each other.

But that way of putting it feels wrong, or bad, or (at any rate) misleading—because any actual physical "bringing-together" of the two particle-1 branches would clearly amount to a violation of the unitarity of the quantum-mechanical equations of motion, and any actual physical "bringing-together" of the two particle-1 branches would clearly amount to a method of transmitting *intelligible signals, instantaneously,* between the α-β-ζ region (on the one hand) and the γ-δ region (on the other), no matter how far apart those two regions might happen to be.

But wait a minute. There is *one* set of circumstances in which the standard textbook presentations of quantum mechanics explicitly *allow* for non-unitary evolutions of the overall quantum state of the world—circumstances (that is) in which *measurements* are occurring, circumstances in which *collapses* are taking place. And *those* (as it happens) are precisely the circumstances in which genuine Bell-type quantum-mechanical *nonlocalities* actually *arise.* And *that* (of course) is not a coincidence.

Let's go back (then) to the state in equation (6), and see if we can think of a way of "bringing together" the two particle-1 branches—not by means of the imposition of some external field (which is impossible), but by means of the performance of a *measurement*. Suppose (in particular) that we measure some observable A, of particle 1, with eigenstates $[A=+1>\equiv [(1/\sqrt{2})[x=\alpha>_1 + (1/\sqrt{2})[x=\beta>_1]$ and $[A=-1>\equiv [(1/\sqrt{2})[x=\alpha>_1 - (1/\sqrt{2})[x=\beta>_1]$.

This kind of "bringing together" of the two particle-1 branches—which all of the standard textbook presentations of quantum mechanics explicitly *allow*—turns out to have very much the same kind of effect on particle 2 as the quantum-mechanically impossible one we considered above! As long as the two particle-1 branches are *kept apart*—as long (that is) as the state in equation (6) obtains—then the two particle-2 branches remain *invisible* to each other. But as soon as the two particle-1 branches are *brought together* by a measurement of A—as soon (that is) as the two particle-1 branches are allowed to *interfere* with each other—then the two *particle-2* branches can interfere with each other *as well*. And what keeps us from *sending messages instantaneously*, in circumstances like these, is the fact that we have no *control* over *which one* of the two possible results of bringing the particle-1 branches together—$[A=+1>$ or $[A=-1>$—is actually going to emerge.

But we are getting ahead of ourselves.

~~~~

Let's go back (then) to the two point-like physical items floating around in a two-dimensional space—and push it one step further. This step (however) will take a bit of setting up.

To begin with, replace the very sharply peaked potential barrier along the diagonal in the two-dimensional space with a more smoothly varying potential *well* along the diagonal in the two-dimensional space. That is: replace the $\delta(x_1-x_2)+\delta(x_3-x_4)$ in (4) with $V(|x_1-x_2|)+V(|x_3-x_4|)$, where $V(|r|)$ is some smoothly varying and monotonically increasing function of $|r|$ that is negative for all finite values of $|r|$ and that asymptotically approaches zero as $|r|$ approaches infinity. This amounts to replacing the sharp contact repulsive force between particles 1 and 2 and particles 3 and 4 in the one-dimensional space with a force that *attracts* 1 *toward* 2 and 3 *toward* 4, and that acts across finite distances (like Newtonian gravitation, say) in the one-dimensional space. And let's stipulate, as well, that this attractive force can be switched on and off as we wish.

The Hamiltonian we're dealing with now (then) is

$$H = (1/2)m_1(dx_1(t)/dt)^2 + (1/2)m_2(dx_2(t)/dt)^2 \\ + (1/2)m_3(dx_3(t)/dt)^2 + (1/2)m_4(x_4(t)/dt)^2 \\ + g(t)(V(|x_1 - x_2|) + V(|x_3 - x_4|)) \\ + \delta(x_1 - x_3)\delta(x_2 - x_4) \qquad (7)$$

where the $g(t)$ is there simply to remind us that we are allowed to switch that part of the potential energy on and off at our discretion.

And note (yet again) that none of these changes are going to alter the fact that the space of ordinary material bodies of a system like this is the one-dimensional space, and that the space of elementary physical determinables is the *two*-dimensional one.

Good. Now suppose (for reasons that will soon become clear) that particles 1 and 3 are much more massive than particles 2 and 4. And set things up as follows: The attractive force is off, and particles 2 and 4 are at rest at the origin, and particle 1 is at the point +1 and particle 3 is at the point −1, as depicted in Figure 1.5.

The corresponding arrangement in the *two*-dimensional space, which is depicted in Figure 1.6, has one of the point-like physical items—item number 1—at the point $(\lambda = +1, \mu = 0)$ and the other—item number 2—at $(\lambda = -1, \mu = 0)$. The arrows in Figures 1.5 and 1.6 indicate the directions in which the two particles (in the case of Figure 1.5) and the two point-like items (in

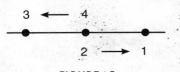

FIGURE I.5

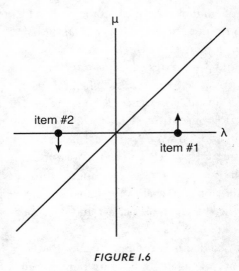

FIGURE I.6

the case of Figure 1.6) will begin to move once the attractive potential is switched on.

Since particles 2 and 4 are *touching* each other here, particles 1 and 3 are going to *bounce off* each other if they should happen to meet. So particles 1 and 3, in *this* situation, do not relate to each other much like components of two distinct *possibilities,* or of two *parallel worlds*. Indeed—and on the contrary—what we are dealing with here (so long as particles 2 and 4 remain at rest, and touching each other, and so long as the attractive forces are switched off) is just the familiar case of two particles (particles 1 and 3) moving around in a one-dimensional space, and interacting with each other by means of a repulsive contact interaction—precisely the case (that is) that we started off with.

And now suppose that we switch the attractive forces on. At this point we will have moved things into a regime in which *both* of the dimensions of the determinable space associated with the Hamiltonian in equation (7) come decisively into play. And one way to think about what's going *on* here is that we have switched on a pair of *measuring devices* for the positions of particles 1 and 3—devices whose *pointers* are particles 2 and 4. When the attractive forces are switched on, particle 2 starts to move in the direction of particle 1, and particle 4 starts to move in the direction of particle 3—particle 2 (you might say) *indicates* the direction in which particle 1 is located, and particle 4 does the same for particle 3—and it was precisely in order to build the appropriate sort of asymmetry into this indicator–indicated relationship that we stipulated, a few paragraphs back, that the masses of particles 1 and 3 be much larger than the masses of particles 2 and 4.

And note (and this is the punch line) that as soon as these measurements take place, and for as long as their different outcomes are preserved in differences between the positions of particles 2 and 4, the whole metaphysical character of the situation—at least as viewed from the one-dimensional material space—appears to radically shift. Any possibility of interaction between particles 1 and 3 is now abolished, and the system behaves, again, for all the world, as if it were a pair of mutually exclusive *scenarios,* or of *parallel universes,* in one of which a light particle detects a heavy particle at position +1, and in the *other* of which the *same* light particle detects the *same* heavy particle, instead, at position −1. And anyone familiar with (say) the many-worlds interpretation of quantum mechanics, or with the decoherent histories interpretation of quantum mechanics, is going to recognize that what we have stumbled across here is precisely the annihilation of the off-diagonal *interference* terms, in the reduced density matrix of a measured system, by the interaction with a *measuring device*—and anyone familiar with *Bohmian mechanics* is going to recognize that what we have stumbled across here is precisely that feature of the linear and unitary evolution of the quantum state that is responsible for the so-called effective collapse of the wave-function.

There is, of course, as yet, no unique determinate fact of the matter about the *outcome* of the sort of measurement described above. What we are left with—once the attractive potential is switched off—is (again) something like a pair of parallel universes, in one of which a light particle has detected a heavy particle at position +1, and in the *other* of which the *same* light particle has detected the *same* heavy particle, instead, at position −1. And the business of arranging for one or the other of those universes to somehow amount to the *actual* one is just the familiar business of solving the quantum-mechanical problem of *measurement:* You either find a way of making one or the other of those universes *disappear* (that's the sort of thing that happens in theories of the "collapse of the wave-function") or you find a way of endowing one or the other of those universes with some sort of special *status* (as one does, for example, in Bohmian mechanics). But we are (again) getting ahead of ourselves. Put the measurement problem to one side for the moment—we'll come back to it later.

———〜〜〜———

We can edge still closer to the familiar quantum-mechanical formalism by replacing the point-like physical items in the space of elementary physical determinables with something more like *fields*. In the examples we considered above, in which different possible situations in the space of ordinary material bodies corresponded to different concrete point-like physical items floating around in the space of elementary physical determinables, the business of arranging for the possibility of *interference* between different such situations involved the introduction of new terms into the fundamental Hamiltonian of the world—terms (for example) like the $\delta(x_1 - x_3)\delta(x_2 - x_4)$ in equations (5) and (7)—whereby the different point-like physical items can literally *push* or *pull* on one another. But fields can do something *else* to one another, something that has *nothing to do* with pushing or pulling, something that doesn't depend on the introduction of any additional terms into the Hamiltonian: they can *add to* or *subtract from* one another—they can *interfere* (that is) in the way that water waves do. So if the inhabitants of the space of elementary physical determinables are something like *fields*, then something like a principle of superposition—then (that is) the possibility of various different possible situations in the space of ordinary material bodies actually physically *interfering* with one another—is going to be built right into the fundamental kinematics of the world, just as it is in quantum mechanics, and it's not going to require any specialized additions to the laws of dynamics.

Here's a toy example.

Go back to the case of a single point-like physical item, floating around in a two-dimensional space, with a diagonal potential barrier. And imagine that we replace that item with a scalar field—a field that always has the value +1 at exactly one of the points in the two-dimensional space, and that always has the value zero everywhere else. And suppose that the point at which the field has the value +1 moves around in the two-dimensional space—just as the point-like physical system did in the earlier example—in accord with the Hamiltonian in equation (3).

It will be natural—just as it was in the case of the single point-like item floating around in a two-dimensional space, and for exactly the same *reasons* as it was in the case of the single point-like item floating around in the two-dimensional space—to describe a world like this as consisting of a pair of ordinary material particles, of different masses, floating around in a homogeneous and isotropic *one*-dimensional space.

Suppose now that there are *two* points in the two-dimensional space at which the field is nonzero, and suppose (just to make things interesting) that the field has the value +1 at one of those points and that it has the value −1 at the other, and suppose that the two points in question move around the two-dimensional space in accord with the Hamiltonian in equation (4). A universe like this one can be described (as before) as consisting of *four* ordinary material particles, moving around in a homogeneous and isotropic one-dimensional space. But the *way* those particles move is (again) kind of funny. If $m_1 = m_3$, and if $m_2 = m_4$, and if we suppose that the signs of the fields in the two-dimensional space make no difference to the intrinsic properties of their one-dimensional projections,[3] then particle 1 and particle 3 are going to be intrinsically *identical* to one another—and yet particle 1 is going to *elastically collide* with particle 2, whereas particle 3 is going to *pass right through* particle 2, and so on. And so (again) a qualitative description of the physical situation at some particular time, in the one-dimensional space, is *not* going to give us enough information to predict, even in principle, the qualitative situation of this world at other times. And so the

---

[3] But "suppose" is really not the right word here. What we are doing here (again) is playing. We are feeling our way. We are making up a theory. And what we are doing, in particular, when we say that the signs of the fields in the two-dimensional space make no difference to the intrinsic properties of the material particles that those fields *correspond* to, is saying something about *what kinds of things* we are imagining these fields to *be*.

two-dimensional space of elementary physical determinables turns out (again) to be more fundamental—in all of the ways that we have already discussed—than the one-dimensional space of ordinary material bodies. And so (again) this turns out to be less like a collection of *four* particles floating around in a one-dimensional physical space than it is like two different *possibilities* about how *two* such particles might be floating around in a one-dimensional space, or like two causally unconnected parallel *worlds,* in each of which a pair of particles is floating around in a one-dimensional space, or something like that.

Except that here—even in the absence of any further modification of the fundamental laws of motion that is designed to enable these two possibilities to *dynamically interact* with each other, even in the absence (that is) of an additional term in the Hamiltonian like the $\delta(x_1-x_3)\delta(x_2-x_4)$ of equations (5) and (7)—they can nevertheless, now and then, and in an altogether different and purely kinematic way, *interfere* with each other. Note (for example) that if the two field-points in the two-dimensional space of elementary physical determinables should ever happen to cross paths, their two fields will cancel each other out. And what that means is that if (for example) the positions of particles 2 and 4 in the *one*-dimensional space of *ordinary material bodies* should ever happen to *coincide,* and if (while the positions of 2 and 4 coincide) particles 1 and 3 should ever happen to come together, then, no matter how far apart particles 1 and 3 may happen to be from particles 2 and 4, *all four of the particles are going to disappear*! This is not the kind of thing (of course) that can happen to nonrelativistic *quantum-mechanical* particles. It would amount (for one thing) to a violation of unitarity. But it is—for all that, and in all sorts of interesting respects—not very far from cases of thoroughly quantum-mechanical interference like the one described in equation (6). And a few obvious further modifications will get us—as literally as you please—all the way there.

~~~~~

Let's start by allowing the field to be nonzero at any number (that is: any *finite* number, or any countably *infinite* number, or any *uncountably* infinite number, or even the entire collection) of points in the space of elementary physical determinables. The different possible physical states of the world (then) will consist of the different possible *configurations* of the *field*—the different possible states of the world (that is) will consist of different possible assignments of field-values to every one of the continuous infinity of points in the space of the elementary physical determinables. And let's suppose that the field $F(\lambda, \mu)$, at any point (λ, μ) in the space of elementary physical determinables, can take on *complex* values, and let's suppose that there is a law (or perhaps an initial condition) to the effect that the integral of $|F(\lambda, \mu)|^2$, over the entirety of the two-dimensional space of determinables, at any particular temporal instant, is equal to 1.

Now, every function of λ and μ that obeys the above stipulations can—as a matter of pure mathematics—be represented as a unique vector, of length 1, in an infinite-dimensional Hilbert space. And we can define—in the familiar way—an inner product on that space. And with that mathematical apparatus in hand, we can stipulate that the evolution of the vector that represents the field-configuration of the world in time is given by some deterministic and linear and unitary time-translation operator whose infinitesimal generator is a Hermitian operator on that space called (by analogy with its classical counterpart) a Hamiltonian.

And now, at last, what we have before us, in all its glory, is what is usually referred to as the *quantum theory* of a pair of nonrelativistic structureless spin-zero particles, floating around, and interacting with each other, in a one-dimensional "physical space." And the business of extending this to N particles, floating around in a three-dimensional space, is now perfectly trivial and familiar. The thing to note (however) is that what this theory presents itself as being *about*, if you come at it by way of the simple and mechanical and flatfooted route that we have been following here, is a *field*. And the space of points on which that field is *defined*, the space of points at which that field takes on *values*, is going to have $3N$ dimensions rather than merely three. And all of the familiar talk of *particles* floating around in a *three*-dimensional space has to do with the way things look from the point of view of the space of *ordinary material bodies*—which (again) is something *other*, and *smaller*, and *less fundamental* than the stage on which the full history of the world plays itself out, and which emerges as a by-product of the action of the Hamiltonian.

~~~~~

This picture of the wave-function as concrete physical *stuff* may seem hard to fit together, at first, with what one thinks one knows about quantum mechanics. Consider (for example) the question of *observables*. We are told—in standard presentations of quantum mechanics—that no measurement can distinguish, with certainty, between a system in the state $F(q_1 \ldots q_N)$ and a system in the state $F'(q_1 \ldots q_N)$ unless the vectors representing $F(q_1 \ldots q_N)$ and $F'(q_1 \ldots q_N)$ happen to be *orthogonal* to each other. We are told (to put it in a slightly different way) that no measurement can distinguish, with certainty, between a system in the state $F(q_1 \ldots q_N)$ and a system in the state $F'(q_1 \ldots q_N)$ unless the vectors representing $F(q_1 \ldots q_N)$ and $F'(q_1 \ldots q_N)$ happen to represent eigenstates, with different eigenvalues, of some Hermetian operator. But why in the world—on this new way of thinking—should anything like that be the case? If these $F(q_1 \ldots q_N)$'s are really concrete physical *stuff*—as opposed to abstract mathematical descriptions of the states of something *else*—why should anything stand in the way of our measuring the *amplitude* of that stuff, to any accuracy we like, at any point we please, just as we are routinely able to do with (say) *electromagnetic* fields?

And this question will be worth examining in some detail, and with some deliberation. The crucial point (as we shall see) is that the very *idea* of measurement is inextricably bound up with the space of *ordinary material things*.

Let's start—a few steps back—with this: The standard textbook presentations of quantum mechanics always include, among the fundamental principles of the theory, among (you might say) the *axioms* of the theory, stipulations (like the one cited above) about what sorts of measurements are possible and what sorts are not. And it has been remarked by any number of investigators that there is something *perverse* about that—it has been remarked by any number of investigators (that is) that claims about what can and what cannot be measured are not the sorts of things that we should expect to find among the fundamental physical principles of the world. A satisfactory set of fundamental physical principles of the world ought to (instead) consist exclusively of stipulations about what there fundamentally *is,* and of laws about how what there fundamentally is *behaves*. And the facts about what can or can not be *measured* ought to *follow* from those fundamental principles as *theorems,* just as the facts about tables and chairs and mosquitoes and grocery stores ought to do. And this (indeed) is the sort of thing that is so entirely taken for granted in the context of classical physics that it is hardly ever so much as remarked upon—and it is the sort of thing that we should demand of any satisfactory proposal for a complete and fundamental science of nature.

Good.

But the business of actually *deducing* theorems about what can and what cannot be measured from some properly constituted set of fundamental principles of physics is obviously going to require that we first have in hand some clear and general and explicit and otherwise serviceable idea of what it is that we are *saying* when we say, of this or that physical quantity, in this or that fundamental physical theory, that it is *measurable*. Consider (then) the following proposal:

> **Proposal M:** The value of a certain physical quantity can be *measured* if and only if there is some physically possible mechanism whereby that value can reliably be *inscribed*, in *ink*, on a *piece of paper*.

This may sound—at first—like a curiously *parochial* analysis of what it is to be "measurable." Why demand (in particular) "ink"? Why demand (in particular) "paper"?

Here's the thought:

1) A physical quantity is said to be *measurable* if and only if nothing at the level of fundamental physical principle stands in the way of the value of that quantity's being *ascertained* by a suitably inclined and sufficiently well-functioning *human being*. This is not at all to suggest that human beings are the only beings that can measure things. The idea is just that it is the doings of human beings that provide a *standard* or a *paradigm* or an *exemplary case* against which other putative cases of "measuring" are to be judged—and (moreover) that being able to measure something involves the possibility, at least in principle, of entering into a certain kind of *relationship* with human beings.

Imagine (for example) that there is a race of superior beings who are able to ascertain the value of a certain exotic physical quantity of which we humans have heretofore had not even so much as an inkling. The very idea that these superior beings are capable of "ascertaining" things suggests that they must *also* be capable, at least in principle, of *communicating with human beings*—because if such communication were somehow fundamentally *out of the question*, then it would be hard to say what it is about these beings that suggests that they are ascertaining (or saying, or thinking, or intending, or measuring) anything at all.

And if (on the other hand) these beings *are* capable of communicating with human beings, and if they are capable of ascertaining the value of this exotic physical quantity, then it follows that we are too—because we can (in principle) ascertain it from *them*.

2) It is an uncontroversial empirical fact of our experience of the world that human beings can more or less read what is written on pieces of paper, and that they can arrange for more or less whatever they like to be *written down* on pieces of paper. And it follows from the fact that human beings can read what is written down on pieces of paper that if there is a physically possible mechanism for reliably writing down the value of a certain physical quantity on a piece of paper, then there is, at least in principle, a way for human beings to *ascertain* what that value *is*. And it follows from the fact that human beings can arrange for more or less whatever they like to *be* written down on a piece of paper that for any physical quantity whose value can even in principle be ascertained by a human being, there is at least one physically possible

mechanism—namely: a suitably inclined and sufficiently well-functioning human being—for reliably *writing* that value *down* on a piece of paper.

And so there turns out to be nothing parochial about Proposal M. We could of course have done just as well by demanding (say) an utterance as opposed to an inscription—or a smoke signal, or a computer file, or any one of an immense variety of physical features of the world that human beings can both observe and manipulate—the crucial point is just that we in no way compromise the *generality* of Proposal M by insisting on any particular *one* of those.[4]

And this sort of reasoning is manifestly going to apply even in cases where the fundamental physical constituents of the world do their to-ing and fro-ing in a mind-bogglingly high-dimensional space of elementary physical determinables, and where the lives of people and drops of ink and pieces of paper are mere *projections* of that to-ing and fro-ing onto the comparatively miniscule three-dimensional space of our everyday experience. All that it takes for granted is that there *are* such things, at one level or another, as people and drops of ink and pieces of paper, and that they behave more or less in the way that we have known, for many thousands of years, that they do.

Good. Let's apply this proposal (then) to *quantum mechanics*.[5]

---

[4] The interested reader can find a discussion of these matters that is both more detailed and more general than the present one in David Z Albert, "The Technique of Significables," in *After Physics* (Cambridge, MA: Harvard University Press, 2015), 89–105.

[5] The next five or six paragraphs are going to require various different kinds of adjustments, at various different points, in the contexts of various different proposals for solving the quantum-mechanical *measurement problem*. But none of that is going to turn out to matter. The reader who is familiar with these proposals—the GRW theory (for example), or Bohmian

In nonrelativistic and non-field-theoretic quantum-mechanical universes, the different points in the space of elementary physical determinables—the different points (that is) at which the quantum-mechanical wave-function *takes on values*—correspond to different configurations of particles in the space of ordinary material bodies. And we have been taking it for granted here that situations in which the configuration of the particles in the space of ordinary material bodies is determinately C are associated with wave-functions whose values are nonzero only at that particular point in the space of the elementary physical determinables that *corresponds* to C. And it follows that the wave-function associated with a situation in which a certain piece of paper bears some determinate inscription and the wave-function associated with a situation in which that same piece of paper bears some *other* determinate inscription are going to have thoroughly *disjoint supports* in the space of elementary physical determinables. And it follows that the wave-function associated with a situation in which a certain piece of paper bears some determinate inscription and the wave-function associated with a situation in which that same piece of paper bears some *other* determinate inscription are going to be perfectly *orthogonal* to each other.

Good.

Now divide the world into three ordinary material systems—a system S whose properties we are interested in measuring, and a piece of paper P, and the rest of the universe R. And consider two distinct possible physical conditions α and β of S. And note that it will follow from Proposal M that α and β can be observationally distinguished from one another if and only if there is at least one physically possible initial condition of P + R on which

---

mechanics, or the many-worlds interpretation—will have no trouble in confirming for herself that none of the adjustments in question are going to affect the overall shape, or the ultimate conclusion, of the argument that follows.

P ends up bearing the inscription "S is in condition α" if S is initially in condition α and on which P ends up bearing the inscription "S is in condition β" if S is initially in condition β.

And note (and this is the crucial point) that it is going to follow from the *unitarity* of the fundamental quantum-mechanical equations of motion—given that the wave-function associated with a situation in which a certain piece of paper bears the inscription "S is in condition α" is orthogonal to the wave-function associated with a situation in which that same piece of paper bears (instead) the inscription "S is in condition β"—that the counterfactual dependence described in the previous sentence can only obtain if the quantum-mechanical wave-function associated any situation in which S is in α is orthogonal to the quantum-mechanical wave-function associated with any situation in which S is in β. And it is a theorem of complex linear algebras that any two orthogonal wave-functions in the same Hilbert space are necessarily both eigenfunctions, with different eigenvalues, of some single Hermetian operator on that space. And that is why the business of observationally distinguishing between any two possible physical conditions of the world must invariably come down to distinguishing between two different eigenvalues of some Hermetian operator, just as the textbooks say.

And from there, without too much further ado, one can recover the entirety of the algebra of the quantum-mechanical observables.

The One-Particle Case

Here's what's happened so far:

We started off by looking at two ways of representing a classical system with two dynamical degrees of freedom, whose Hamiltonian consists of the standard kinetic energy terms and a simple contact interaction. One of these represents the system as a pair of particles floating around in a one-dimensional space of ordinary material bodies, and the other represents the system by means of a single point-like physical item in a two-dimensional space—the space of the possible *one*-dimensional *configurations* of the pair of particles floating around in the space of ordinary material bodies. Because these two ways of representing the system are both complete, and because they are fully isomorphic to each other, and because the two-dimensional representation looks (in all sorts of ways) less natural, and less familiar, and less like the manifest image of the world than the one-dimensional representation does, there seemed to be no compelling reason to take the two-dimensional space philosophically seriously.

But as soon as we imagine an *additional* point-like physical item floating around in the two-dimensional space, all of this abruptly changes. Once the two-dimensional space is inhabited by more than a single such item, the one-dimensional representation and the two-dimensional representation are no longer mathematically equivalent to one another—and *each* of them seems to have a distinct and philosophically interesting role to play. The one-dimensional space is still the space of ordinary material bodies—but the representation of the system in that space is no longer mathematically complete. And the smallest space in which the system can be represented in a complete and separable way[6]—the space (that is) of the elementary physical determinables—is now *two*-dimensional.

Moreover, the general *direction* of these changes is unmistakably *quantum-mechanical*. It turns out that adding another concrete point-like fundamental physical item to the

---

[6] What it means for a representation to be *separable*, by the way, is for that representation to take the form of a spatial distribution of local physical properties. What it means (that is) for a representation to be separable, in the language I introduced a few pages back, is for it not to involve any *entanglement*.

higher-dimensional space is not so much like adding more *concrete physical material* to the lower-dimensional space as it is like adding another low-dimensional *world*, or another somehow also actualized low-dimensional *possibility*, or (better, and more precisely) another *term* in a quantum-mechanical *superposition*. And these different possibilities can be made to *interact* with one another, in ways that are very much reminiscent of quantum-mechanical *interference*, by means of the addition of another very simple term to the Hamiltonian. And the addition of such a term also generates distinctly quantum-mechanical sorts of nonlocality, and distinctly quantum-mechanical images of measurement, and so on.

And (on top of that) it happens to be a characteristic of *classical* physical theories that the space of ordinary material things and the space of elementary physical determinables are exactly and invariably *one and the same*. And it seems natural to wonder whether all of this points to some kind of a *diagnosis*, or some kind of an *explanation*, of the actual un-classical weirdness of the world. It seems natural to wonder (that is) whether it is precisely this coming-apart of the space of ordinary material things and the space of elementary physical determinables that turns out to be at the bottom of everything that's exceedingly and paradigmatically *strange* about quantum mechanics.

But how (it is time to ask) can that possibly be true? For quantum-mechanical systems consisting of just a single structureless spin-zero particle (after all) the space of ordinary material bodies and the space of elementary physical determinables are precisely one and the same, just as they are for classical systems. And yet a hell of a lot of what everybody agrees is exceedingly and paradigmatically strange about quantum mechanics can already be encountered in systems like that.[7] And this will

---

[7] Richard Feynman famously says (for example) that the *only* mystery in quantum mechanics is the one that comes up in connection with the

be worth thinking through in some detail. And the business of thinking it through will be the work of this section.

Consider (then) a single structureless particle, in a three-dimensional space of ordinary material things, whose quantum-mechanical wave-function happens to be nonzero, at a certain particular time, in two separate and compact and disjoint regions of that space called A and B.

What's strange about situations like this is that both of the following claims about the particle in question are apparently, simultaneously, true:

1) There is a perfectly concrete and observable sense in which the particle, or something very closely associated with the particle, is in *both* regions. (What I have in mind here, when I speak of a "concrete and observable" sense in which the particle is in both regions, is of course the possibility of measuring the effects of *interference* between the branch of the wave-function that's located in A and the branch of the wave-function that's located in B—as, for example, in the double-slit experiment)

2) There is a perfectly concrete and observable sense in which the particle, and everything sufficiently closely associated with the particle, is in only *one* of those regions. (And what I have in mind here, when I speak of a "concrete and observable" sense in which the particle is in only one region, is the fact that if we measure the particle's spatial location, either we will find a particle in

---

double-slit experiment—and the double-slit experiment seems (on the face of it) to involve nothing over and above a single structureless particle moving around in the presence of a complicated (double-slitted) external potential. See Richard P. Feynman, Robert B. Leighton, and Matthew Sands, *The Feynman Lectures on Physics,* vol. 3 (Reading, MA: Addison-Wesley, 1965), chap. 1.

A and nothing whatever in B, or we will find a particle in
B and nothing whatever in A)

Note (to begin with) that there is nothing particularly unintelligible, in and of itself, about claim (1). (1) is what Bohr and his circle used to call the "wave" aspect of quantum-mechanical particles—and one could think of that, in the absence of (2), as suggesting a novel but by no means unfathomable picture of the subatomic structure of matter, according to which particles are to be understood, at the microscopic level, as something akin to *clouds* or *fluids* or *fields* that can (in certain circumstances) spread themselves out over finite and even disjoint regions of the space of ordinary material things.

What has always completely *freaked everybody out* (on the other hand) is the combination of (1) *and* (2). And it turns out that all of the ways that we have of imagining that (1) and (2) could (somehow) *both* be true are going to involve telling stories about systems that consist of *more* than a single particle, systems (that is) whose quantum-mechanical wave-functions take on values at points in spaces of more than three dimensions, systems (that is) for which the space of elementary physical determinables *diverges* from the space of ordinary material things.

Let me try to explain, a little more concretely, what I have in mind.

Note (to begin with) that the business of figuring out how (1) and (2) could both be true is nothing other than the business of solving the quantum-mechanical measurement problem. And so the various attempts at coming to terms with (1) and (2) together that we ought to have in the back of our minds here are things like the GRW theory, and Bohmian mechanics, and the many-worlds interpretation. And it turns out that all of those attempts, and all of the strategies that anybody has ever so much as hinted at for solving the quantum-mechanical measurement problem, depend on the phenomenon of *entanglement*. And the phenomenon of entanglement is, as we have noted before, and as a straightforward matter of definition, the phenomenon of the divergence of the space of elementary physical determinables from the space of ordinary material things.

Consider (for example) the case of Bohmian mechanics. The phenomena that pertain to (1) have to do—in the context of Bohmian mechanics—with the fact that the wave-function of the sort of particle we were talking about above is nonzero both in region A and in region B. And it might appear, at first glance, as if the business of accounting for the phenomena that pertain to *(2)*, in the context of Bohmian mechanics, amounts to nothing more than the simple observation that—notwithstanding that the wave-function of this particle is spread over both region A and region B—the Bohmian corpuscle itself is located either in region A or in region B. And all of that can of course be presented in the form of a story about what things are physically like, at various different times, at various different points in the familiar three-dimensional Euclidian space of ordinary material things, and of our everyday empirical experience of the world.

But this (on a little reflection) is all wrong. What (2) is about is not merely that the particle *is* either in region A or in region B, but (in addition) that when we *look* for the particle, we *see* it either in region A or in region B, and (moreover) that what we see is in fact reliably correlated with where the particle actually *is*, and that what we see matches up in the appropriate way with what we would see if we were to look *again*, and with what somebody *else* would see if they were to look for *themselves, and with how the particle itself will behave in the future,* and so on. And if not for all that, there would (indeed) be nothing here to puzzle over. And the various businesses of accounting for all that, in the context of Bohmian mechanics, all depend (again) on the fact that the process of measurement invariably and in-

52   A GUESS AT THE RIDDLE

eluctably generates quantum-mechanical entanglements between the measuring devices and the measured particle.[8]

Let's have a look at exactly how that works. Start with a single, structureless, particle—call it p—and two boxes. One of the boxes is called A, and is located at the point ($x=+1, y=0, z=0$), and the other is called B, and is located at the point ($x=-1$,

---

[8] Maybe it will be worth taking a minute to rub this in.

Suppose that the initial wave-function of the composite system consisting of a particle (p) and a measuring device (d), which is designed to record the position of that particle, is:

$$[\text{ready}>_d(\alpha[A>_p+\beta[B>_p), \qquad (i)$$

where [ready>$_d$ is the physical state of the system d in which d is plugged in and properly calibrated and facing in the right direction and in all other respects ready to carry out the measurement of the position of p, and [A>$_p$ is the state of p in which p is localized in the spatial region A, and [B>$_p$ is the state of p in which p is localized in the spatial region B.

And note, to begin with, that any satisfactory scientific account of why it is that if we measure the position of a particle like this "we will either find a particle in A and nothing whatever in B, or a particle in B and nothing whatever in A" has got to be an account, not only of the behavior of p under circumstances like (i), but also of the behavior of d under circumstances like (i).

Good. Suppose that p and d are allowed to interact with one another, in the familiar way, when a state like (i) obtains. Then it will follow, in the familiar way, from the linearity of the quantum-mechanical equations of motion, and from the stipulation that d is a properly functioning device for the measurement and recording of the position of p, that the state of this composite system once this interaction is complete will be:

$$\alpha['A'>_d[A>_p+\beta['B'>_d[B>_p, \qquad (ii)$$

where ['A'>$_d$ is the state of d in which the position of d's pointer indicates that the outcome of the measurement of the position of p is 'A,' and ['B'>$_d$ is the state of d in which the position of d's pointer indicates that the outcome of the measurement of the position of p is 'B.'

And consider how Bohmian mechanics manages to guarantee that, in circumstances like (ii), the positions of the Bohmian corpuscles that make up the pointer of d are properly and reliably correlated with the position of the Bohmian corpuscle p—the position (that is) of the Bohmian corpuscle whose position has just now been *measured*. Note (in particular) that that correlation depends crucially on the fact that the wave-function in (ii) vanishes in those regions of the configuration space of the composite system consisting of p and d in which precisely those correlations do not obtain. And note (more-

$y = 0$, $z = 0$).[9] And let $[A>_p$ be the state of p in which p is located in A. And let $[B>_p$ be the state of p in which p is located in B. And suppose that at $t = 0$, the state of p is

$$(1/\sqrt{2})[A>_p + (1/\sqrt{2})[B>_p. \qquad (8)$$

Now, states like the one in (8) famously resist any interpretation as situations in which p is either in box A or in box B. And what famously *stands in the way* of such an interpretation is the fact that if we open both boxes, when a state like (8) obtains, then the subsequent observable behaviors of the particle—the probabilities (for example) of finding the particle at this or that point in space—are in general going to be very different from the behavior of a particle released from box A, and very different (as well) from the behavior of a particle released from box B, and very different (as well) from anything along the lines of a probabilistic sum or average of those two behaviors.

And all of this, as I mentioned above, can be explained, in the context of Bohmian mechanics, by means of a story about what things are physically like, at various different times, at various different points in the familiar, material, three-dimensional space of

---

over) that it must vanish in those regions without vanishing throughout those regions of that configuration space in which p is located in A, and without vanishing throughout those regions of that configuration space in which p is located in B, and without vanishing throughout those regions of that configuration space in which d's pointer is located in 'A,' and without vanishing throughout those regions of that configuration space in which d's pointer is located in 'B.' And note (and this, finally, is the heart of the matter) that the previous two sentences can simultaneously be true only of a wave-function (like the one in (ii)) in which p and d are quantum-mechanically *entangled* with one another—note (that is) that the previous two sentences can simultaneously be true only of a wave-function that (like the one in (ii)) cannot be represented as a function over the points of any three-dimensional arena.

[9] In order to keep things as simple as possible, we will treat these boxes not as physical systems, but (instead) as externally imposed potentials—and we will treat the openings and closings of those boxes not as dynamical *processes,* but (instead) as *variations* in those externally imposed potentials with time.

our everyday empirical experience of the world. The particle itself starts out in either box A or box B—but its wave-function, its so-called pilot-wave, is nonzero (when a state like (8) obtains) in both boxes. And so, when the boxes are opened, and the two branches of the wave-function flow outward, and fill up the three-dimensional space around them, and overlap with one another, they interfere—and that interference observably affects the motion of the particle that those two branches, together, are guiding.

And the puzzle (again) is that measuring the position of a particle like that somehow makes one or the other of those branches go away. And the question is *how*. The question (to put it as naively and as literally and as flatfootedly as one can) is *where*, exactly, does that other branch *go*?

Consider (then) a radically simplified stand-in for a measuring device—call it M—that consists (just as in the case we considered before) of a single structureless particle, and which is constrained (we will suppose) to move along the X-axis. And let [ready>$_M$ be the state of M in which M is located at the point ($x=0$, $y=0$, $z=0$), and let ['A'>$_M$ be the state of M in which M is located at ($x=+1/2$, $y=0$, $z=0$), and let ['B'>$_M$ be the state of M in which M is located at ($x=-1/2$, $y=0$, $z=0$). And suppose that the kinetic term in the Hamiltonian of M happens to be identically zero. And suppose that there is an interaction between M and p that we can switch "on" and "off" as we please, and that (when it's switched "on") produces (over the course of, say, the ensuing second) evolutions like this:[10]

$$[\text{ready}>_M[A>_p \to ['A'>_M[A>_p \text{ and}$$
$$[\text{ready}>_M[B>_p \to ['B'>_M[B>_p \tag{9}$$

---

[10] Here again, just to keep things simple, we are going to treat the business of turning this interaction "on" and "off" not as a variation in any dynamical degree of freedom, but (instead) as a variation in an externally imposed effective Hamiltonian. None of these simplifications—as the reader can easily confirm for herself—involves any loss in generality.

When the interaction is switched "on" (then) M functions as a measuring instrument for the position of p.

Note that whereas the space of the elementary physical determinables of a single structureless quantum-mechanical particle p is *three*-dimensional, the space of the elementary physical determinables of the *composite* quantum-mechanical system consisting of p and *M*—call that $\mathbf{F}_{pM}$—is going to be *four*-dimensional. We can assign unique addresses to points in that arena using the three coordinates (call them x_p, y_p, and z_p) that correspond to the parochial three-dimensional "position of p" and one more (call it x_M) that corresponds to the parochial one-dimensional "position of M."

Suppose that we initially that prepare this composite system in the state

$$[\text{ready} >_M ((1/\sqrt{2})[A >_p + (1/\sqrt{2}) [B >_p), \qquad (10)$$

with the interaction switched "off," and then open the boxes. In this case, the M remains completely unentangled with p, and once the boxes are opened, one branch of the wave-function of the composite system will spread outward from the point ($x_p = +1$, $y_p = 0$, $z_p = 0$, $x_M = 0$), and the other branch will spread outward from the point ($x_p = -1$, $y_p = 0$, $z_p = 0$, $x_M = 0$), and each of them will fill up the three-dimensional hypersurface $x_M = 0$ of the determinable space of the composite system, and they will overlap with each other, and they will interfere with each other, and both of them will contribute to determining the Bohmian trajectory of the world-particle. (And note that all this—except for the presence of the world-particle itself—is exactly analogous to what was going on in the system described by the Hamiltonian in equation (7) when particles 2 and 4 are both at the origin)

If (on the other hand) we initially prepare the composite system in the state in equation (8) with the interaction switched "on," then it will follow from (9), together with the linearity of the quantum-mechanical equations of motion, that the state of p + M will become

$$(1/\sqrt{2})['A'>_M[A>_p + (1/\sqrt{2})['B'>_M[B>_p \quad (11)$$

Now M and p are maximally entangled with one another, and if the boxes are opened at this point, then one branch of the wave-function of the composite system will spread outward from the point ($x_p = +1$, $y_p = 0$, $z_p = 0$, $x_M = +1/2$), and fill up the three-dimensional hypersurface $x_M = +1/2$, and the other branch will spread outward from the point ($x_p = -1$, $y_p = 0$, $z_p = 0$, $x_M = -1/2$), and fill up the three-dimensional hypersurface $x_M = -1/2$, and the two will not overlap with each other, and will *not* interfere with each other, and only one of them—the one that's nonzero on the hypersurface where the world-particle happens to be located—will contribute to determining the trajectory. And the reader should note that it is absolutely critical to the way in which all this works—it is absolutely critical (in particular) to the very idea of an *entangling* of the measuring device with the measured particle—that the dimension of the determinable space along which the wave-function spreads out when *M* is in motion is *orthogonal* to all of the dimensions of that space in which the wave-function spreads out when *p* is in motion. (And note that *this*—except (again) for the presence of the world-particle itself—is exactly analogous to what was going on in the system described by the Hamiltonian in equation (7) when the attractive potential is switched on)

And so the answer to the question of where the other branch goes, when we measure the position of p, is literally, and flat-footedly, that *it gets pushed off into another dimension*. And this (in microcosm) is the sort of thing that happens *whenever* we perform measurements on quantum-mechanical systems. What's strange about quantum mechanics, what makes it look like magic, even in a case as simple as that of a single structureless particle, is that the three-dimensional space of ordinary material bodies is too small to contain the complete microscopic history of the world.

And much the same sort of thing is true on the GRW theory. This may seem, at first, like a puzzling claim. The reader might want to object that what happens on the GRW theory is not that one of the branches gets "pushed off into another dimension," but (instead) that one of the branches simply *disappears*. But consider the *mechanism* of that disappearance. The wave-function of the world, which is a function of position in the space of elementary physical determinables, is multiplied by another function, the so-called hitting function—which is *also* a function of position in the space of elementary physical determinables. And this multiplication of the wave-function by the hitting function somehow manages to leave one of the above branches of the wave-function intact, and causes the other one to vanish. And that can only occur if these two branches of the wave-function, which overlap everywhere in the three-dimensional space of ordinary material bodies, somehow manage *not* to overlap *anywhere* in the space of elementary physical determinables. And *that* can only occur if the space of elementary physical determinables has at least one more dimension than the space of ordinary material bodies—and if the two branches have somehow become separated from each other along that additional dimension.

And the reader can confirm for herself that much the same thing would be true, as well, on the many-worlds interpretation of quantum mechanics—if the many-worlds interpretation were not otherwise incoherent.[11]

---

[11] The incoherence I have in mind here has to do with the business of trying to find a place for quantum-mechanical *probabilities* in the many-worlds interpretation of quantum mechanics.

And so, at the end of the day, there *does* seem to be an intimate and invariable connection between the coming-apart of the space of ordinary material things and the space of elementary physical determinables (on the one hand) and everything that's exceedingly and paradigmatically *strange* about quantum mechanics (on the other). Quantum-mechanical sorts of behavior seem to *require* that the space of the elementary physical determinables is bigger than the space of ordinary material things—and whenever the space of the elementary physical determinables space *is* bigger than the space of ordinary material things, quantum-mechanical sorts of behavior seem to quickly ensue. And it begins to look as if what we have stumbled across here is (indeed) a *diagnosis,* or an *explanation,* of the fact that the world is quantum-mechanical.

Conclusion

Let's see where we are.

The fact that the space of the elementary physical determinables of the world and the space of the ordinary material bodies of the world are conceptually *distinct* from each other—the fact that there is absolutely no a priori reason they should *coincide* with each other, or have the same *topology* as each other, or have the same *dimensionality* as each other—is a purely *logical* point, a point that might in principle have been noticed, by means of purely conceptual analysis, long before the empirical discoveries that gave rise to quantum mechanics. And we have seen how easy it is, merely by playing around with the simplest imaginable Hamiltonians of classical Newtonian particles, to stumble onto physical systems for which the space of the elementary physical determinables has a different number of dimensions than the space of ordinary material bodies. But (as I have already remarked) there is nothing mysterious or surprising about this distinction's having in fact gone unnoticed for as long as it did. It is (after all) a fundamental principle of the Manifest Image of the World—and *all the more so* (indeed) because we are not even aware of ever actually having *adopted* it—that the material space of the world and the determinable space of the world are exactly the same thing. And that principle has since been endorsed, and further fortified, in the course of scientific investigation, by Newtonian mechanics, and by Maxwellian electrodynamics, and by the special and general theories of relativity, and even (insofar as these can be considered in isolation from quantum mechanics) by the high-dimensional geometries of string theory, and (indeed) by the entire edifice of classical physics. You might even say that the principle that the material space of the world and the determinable space of the world are exactly the same thing is the very *essence* of the classical picture of the world, and the simplest and most illuminating way of pointing to what sets it apart from quantum mechanics.

But the relationship between the material space and the determinable one is (for all that) a *contingent* matter. And one of the lessons of the simple exercises we have been working our way through here is that *the moment that we take that in,* the moment that we even *raise the question* of what the world might be like if those two spaces *differed* from one another, something paradigmatically quantum-mechanical just flops right out. And it seems fair to say that if the conceptual distinction between the material space of the world and the determinable space of the world had made itself clear to anybody (say) 150 years ago, then the twentieth-century physics of subatomic particles might have amounted to less of a shock than, in fact, it did—it seems fair to say (that is) that the *clarification* of the conceptual distinction between the material space of the world and the determinable space of the world offers us a way of looking at quantum mechanics as something natural, and beautiful, and simple, and understandable, and maybe even to be expected. Indeed, in the light of the sorts of considerations that we have been through here—the *classical* case is the one that looks exceptional, and conspiratorial, and surprising.

~~~~~

The trick (in a nutshell) is to learn to think of what we see in our experiments as the to-ings and fro-ings of *shadows on a wall*. The idea is that all of those to-ings and fro-ings can actually be *understood*, the idea is that all of those to-ings and fro-ings can actually be *explained*, notwithstanding their almost unspeakable original weirdness, in terms of a simple and visualizable and mechanical account of something that's going on (as it were) *behind our backs*, in a part of the cave at which we are forbidden to directly look.

2

Physical Laws and Physical Things

Recent discussions of the foundations of quantum mechanics have raised what seem to me to be genuinely new and puzzling and urgent questions about exactly what it is that distinguishes the category of the nomic from the category of the concrete. And what I want to do here is to put some of those questions, as explicitly and as provocatively as I can, on the table.

Let me start out by talking a bit about the old project of writing down a relationalist version of Newtonian particle mechanics. This project has a long and complicated history, as everybody knows, that stretches back over hundreds of years, to the beginnings of Newtonian mechanics itself. And there is much too much of it to say anything comprehensive about it here. Let me just remind you of two of its highlights.

There is a Machian proposal that the inertial frames of reference—the frames of reference (that is) with respect to which $F = ma$ is a true law of nature—are the ones that are un-accelerated and un-rotating with respect to what people used to call the "bulk mass" of the universe.

And there is a beautiful and mathematically ingenious proposal due to Barbour, in which the physically possible trajectories of the world turn out to be geodesics in what he calls a "shape" space, whose points correspond to sets of interparticle distances.

Both of these theories deny, as any properly relationalist theory should, that there is ever any fact of the matter about where anything is, or about how anything is moving, with respect to any background, absolute space, or with respect to anything other than other material bodies. And both of them deny, as any properly relationalist theory should, that there is any dynamical law that would apply to the case of a single material body alone in the universe.

But both of them also have features that one might not initially think one *wants* in such a theory. Neither of them reproduces *all* of the purely relational consequences of Newtonian mechanics, and both of them include a curious kind of nonlocality.

Consider, for example, the case of a system (call it S) that consists of two masses connected by a spring. And suppose that at a certain particular moment the distance between the two masses is somewhat greater than the relaxation length of the spring, and that the first derivative of that distance, with respect to time, is zero.

In an absolutist Newtonian mechanics, there are any number of different ways in which the distance between the masses might evolve with time. That distance might remain constant (which is what would happen if the system were rotating about its center, at the appropriate angular velocity, with respect to the background absolute space), or it might oscillate (which is what would happen if the rotation with respect to the absolute space were slower than in the previous case, or if it were zero).

But in either of the relationalist theories I mentioned above, what will happen to a system like this is going to depend on what *other* systems there might or might not happen to be in the world we are imagining—even if those other systems happen to be arbitrarily far away from S, and even if they happen to have no dynamical interaction with S.[1] Both of those theories entail, for example, that a system like S, if it is all alone in the universe, can *only* oscillate—since, in that case, the inertial frames are going to have to be ones in which S is un-rotating—and all of the continuous infinity of *other* substantivalist Newtonian solutions simply *disappear*.

But there is an astonishingly simple and obvious and trivial way of writing down a relationalist version of Newtonian particle mechanics that has all of the advantages, and none of the disadvantages, of the two theories we have just been talking about, and that can be stated, in its entirety, as follows: **The physically possible sequences of interparticle distances are just the ones that have at least one faithful embedding into an Aristotelian space-time such that, on that embedding, they satisfy $F = ma$.**

This theory (let's call it the *embedding* theory) recovers the entirety of the relational consequences of the substantivalist version of Newtonian mechanics and *nothing else*. It has nothing like the nonlocality of the theories we considered above. It is a theory in which any dynamically isolated system can be treated entirely on its own, as if it were alone in the universe, just as we are used to in the absolutist version. And so on.

What's the trick here? How do we end up *getting rid* of space as a *substance*? What happens is that we *absorb* the space into the category of *law*.

And what's *wrong* with this? Why haven't people *availed* themselves of this? Why has everybody been knocking themselves out, for so many hundreds of years, wasting their energy and their brilliance, torturing themselves with compromises, when there was an utterly *obvious* and *effortless* and *trivial* move

---

[1] By which I mean: even if those other systems exert no *forces* on S, and exchange no *energy* with S.

that could have gotten them absolutely everything they ever could have wanted?

~~~~

Well, let's see. There is a persistent accusation in the literature to the effect that this theory is somehow *disingenuous*—that it *uses* the substantival Aristotelian space-time, that it *relies* on the substantival Aristotelian space-time, and then turns around and dismisses it as a fiction.

But that just doesn't seem right to me. I don't see how the geometrical structure of Aristotelian space is ever being treated here as if it were the structure of some substantial *thing*—it seems to me that it is being treated, from beginning to end, as an exquisitely concise and efficient *mathematical device* for expressing the content of the *purely relational laws*. Suppose (for example) that I have a theory according to which the trajectories of celestial bodies—or maybe the projections of those trajectories onto a certain two-dimensional surface—exactly coincide with the lines that would be traced out by a certain particular kind of perfectly rigid galaxy-sized spirograph. And suppose (moreover) that I *describe* my theory, that I *present* my theory, in precisely those terms. Is a theory like that—or is this particular *presentation* of a theory like that—somehow disingenuously helping itself to perfectly rigid galaxy-sized spirographs only to turn around and dismiss them as a fiction? Of course not! The theory is merely availing itself of a compact and convenient device for expressing what it has to say about the motions of the celestial bodies. And it seems to me that very much the same sort of thing is going on in the embedding theory.[2]

---

[2] It might be objected that the kinds of *explanations* that this embedding theory has to offer us are somehow *second-rate*—it might be objected (in particular) that the kinds of explanations that this embedding theory offers us are *extrinsic* explanations rather than *intrinsic* ones, in the sense of extrinsic /

But there are more serious worries that can be raised about a theory like this. Some people—Cian Dorr, for example—have worried that the trick here—the business (that is) of *absorbing* whatever it is that we don't want in the *concrete physical ontology* into the category of the *nomic*—is too easy. The worry (in particular) is that you can do this, once you get the hang of it, with *anything*.

~~~~

Suppose (for example) that you would like to remove some particular particle, or that you would like to remove *all* of the particles except for the ones *inside your head*, or that you would like to remove whatever particular particles you happen not to *like*, from the universal catalogue of concrete material things.

Suppose (that is) that you would like to remove some particular M of the particles mentioned in the standard Newtonian theory of the world—any M of those particles that you like—and leave only the remaining N.

Here's how to write down the Newtonian theory of the motions of the N particles that you have decided to hang on to: **The physically possible motions of the N existing particles are just the ones that have at least one faithful embedding into a *larger* collection of $N+M$ particles, floating around in background Aristotelian absolute space-time, in accord with $F=ma$.**

---

intrinsic that was first introduced in Hartry H. Field, *Science without Numbers: A Defence of Nominalism* (Princeton, NJ: Princeton University Press, 1980). But *that* is going to count as an objection to any theory whose correct logical formulation quantifies over *any abstract objects at all*—and it is going to count (in particular) as an objection to any theory whose correct logical formulation quantifies over *numbers*—and so it is certainly *not* going to count as an objection to the embedding theory *in particular*, or to the embedding theory *as opposed* to *Mach's* theory, or as opposed to *Barbour's* theory, or as opposed to old-fashioned absolutist Newtonian mechanics.

What's going on *here* is that we have absorbed the *M particles* that we don't like into the *laws* of the motions of the *N* particles that we *do* like—just as we absorbed the *space* that we don't like into the laws of the evolutions of the *interparticle distances* that we *do* like in the example above.

And now it begins to look as if this sort of a trick is going to allow you to get rid of *anything* you don't happen to *like*. And this is what seems to be worrying people like Cian. But I think this is a bit too fast.

The theory we have been entertaining just now is—in all sorts of ways, and in stark contrast to the relationalist theory that we were talking about before—a mess, and a failure, and a joke. Note (to begin with) that the relationalist theory that we were talking about before is *deterministic*. On *that* theory, the sequence of interparticle distances over any arbitrarily short interval uniquely determines the entire sequence, out to $t = +$ and— infinity. But the present theory—the one (that is) about the *N* particles that we happen to like—gives us no lawlike connections *at all* (not deterministic ones, and not chancy ones either) between the positions of the *N* existent particles at one time and their positions and velocities at some other time, or between the positions of the particles at one time and their positions throughout some (nonoverlapping) finite *interval* of time, or between the positions of the particles at one time and their positions throughout some (nonoverlapping) *semi-infinite* interval of time! *Nothing!*

Here's an easy way to see that: Imagine that one of the *M* fake particles never interacts with any of the *N* real particles until a certain time t_c—at which point it collides with real particle #14, altering particle 14's direction of motion. In that case, nothing in the motions of the *N* real particles, all the way from $t = -$infinity to t_c, is going to offer any hint at all of what that fake particle is doing, or when or how or whether in might be expected to interfere with the motions of any of the *real* particles. And so there is clearly going to be a continuous infinity of *different* trajectories of the *N* real particles, all of which are fully in accord with the laws of this theory, and all of which *exactly and entirely coincide with one another* for all times prior to $t = t_c$.

And note that it is not going to do so much as an iota of good, at this point, to relent, and to allow one or two or three of the banished *M* particles back into the catalogue of concrete physical things. Letting any proper subset of them back in is going to do nothing whatsoever—but the minute you let *all* of them back in—then: *poof!*—you have a recognizably dynamical theory. And not (mind you) just *any* recognizably dynamical theory—but the *best* and most *informative* and most *explanatory* kind of recognizable dynamical theory: a fully *deterministic* theory! And so it feels as if the world is palpably *screaming* at you here that *all* of the particles are concrete physical things, and that the business of trying to absorb so much as a single one of them into the domain of the *nomic* is a simple *mistake*. And nothing like that goes on in the case of the analogous relationalist strategy that we were talking about before.

~~~~~

I have been supposing, just to keep things simple, that the interactions between these "particles" (the interactions—that is—between the "real" particles and the other "real" particles, and between the "real" particles and the "fake" particles, and between the "fake" particles and the other "fake" particles) are all what physicists call *contact* interactions. I have been supposing (in other words) that these "particles" exert forces on one another only when they actually *smack into* one another—like (say) rocks, or billiard balls. But what if that's not the case? What would happen (for example) if the "particles" all interacted with one another by means of old-fashioned, infinite-range, instantaneous, Newtonian *gravitational* forces?

Let's see.

One thing to say is that the special theory of relativity is simply not going to *accommodate* forces like that. And (come to think of it) the special theory of relativity is not going to accommodate *any* interaction that might somehow weaken or diminish or delimit the upshot of the argument I presented above—the argument (that is) that took it for granted that all of the interactions between these "particles" were *contact* interactions. Here's what I mean: Suppose that the trajectories of every one of the N "real" particles passes through some space-time region R. And suppose that Q is some space-time region that lies entirely outside of the past light-cone of R. In that case, nothing about the motions of the N real particles anywhere in the absolute past of R can offer any hint of what the *fake* particles might or might not be up to in region Q, or of when or how or whether any of those M fake particles might be expected to *interfere*, in the absolute *future* of R, with the motions of any of the N *real* ones. And so there is clearly going to be a continuous infinity of *different* trajectories of the N real particles, all of which are fully in accord with the laws of this theory, and all of which *exactly and entirely coincide with one another* throughout the absolute past of R. And so there can be no Lorentz-invariant laws *whatsoever* of the motions of the N real particles—not deterministic ones and not chancy ones either.

And that—for all practical purposes, for all physically realistic purposes—is that. But it turns out that there is something useful to be learned by pressing this worry just a little further.

Suppose (then) that we are dealing with a physically impossible nonrelativistic Newtonian universe in which the various "particles" in our scenario have instantaneous infinite-range interactions with one another. And suppose (moreover) that enough information about the initial positions and velocities of the M "fake" particles somehow manages to get *inscribed*, by means of those interactions, onto the initial motions of the N "real" particles, so that the *initial* motions of the N "real" particles suffice to uniquely determine the *later* motions of the N "real" particles, so that (in other words) the "embedding" strategy we have been talking about here produces a fully deterministic theory of the motions of the N "real" particles *all by themselves*.[3]

Let's consider a concrete case. Suppose that our $N+M$ "particles" are something like billiard balls. Suppose (that is) that they interact with one another both by means of electrostatic contact interactions *and* by means of long-range gravitational interactions—and suppose, as well, and (again) as with *actual* billiard-balls, that the strength of the long-range gravitational interaction is tiny compared with the strength of the short-range electrostatic interaction. In a case like that, the resulting deterministic theory of the motions of the N "real" particles is going to be one in which the later motions of the real particles depend in astronomically sensitive ways on their earlier motions. Indeed, as the strength of the gravitational interaction approaches zero, this sensitivity will tend toward infinity—and at the limit we will be returned to precisely the kind of radical indeterminism that we were discussing above. And note that the instant that we admit all of the "fake" particles into the catalogue of what there is—*poof*—all of these pathologies disappear. The dynamics is revealed as something that connects the earlier and later motions, not of the N particles that we initially

---

[3] Whether such an inscribing is actually *possible*—by means of anything like (say) instantaneous Newtonian gravitational interactions between the "particles"—is (by the way) a very difficult mathematical question. But the point here is to make the worry as acute as we can. Suppose (then) that such an inscribing *is*, actually, possible.

preferred, but of nothing short of the full set of $N+M$ particles. And the sensitivity to initial conditions evaporates, and the gravitational interactions—and even the very question of whether or not there *are* any gravitational interactions—are exposed as essentially irrelevant to the question of how these particles actually *move*. And (again) it feels as if the world is palpably *screaming* that *all* of the particles are concrete physical things, and that the business of trying to absorb so much as a single one of them into the domain of the *nomic* is a simple *mistake*. And (again) nothing like that goes on in the case of the analogous relationalist strategy that we were talking about before.

~~~~~

Ok. Let's try a slightly different tack. What we want, again, is to find some way of absorbing the M particles that we don't like into the laws of the motions of the N particles that we do like. How (then) about this: **The physically possible motions of the N existing particles are just the ones that have at least one faithful embedding into a *larger* collection of $N+M$ particles—*where the positions and velocities of the M auxiliary particles, at some specified time t, are stipulated to be* $(x_1 \ldots x_{3M}, v_1 \ldots v_{3M})$—floating around in background Aristotelian absolute spacetime, in accord with $F=ma$.**

In *this* theory—unlike in the previous one—the law makes explicit reference to some *particular* set of initial positions and velocities for the M "imaginary" particles. So *this* theory is not going to suffer from any of the pathologies of the previous two.[4] But this theory is going to feature a fantastically *complicated* law of the motions of the N "real" particles. So the theory in question here is going to be perverse in a different sense.

---

[4] The one (that is) with only contact interactions, and the one with the particles that interact like billiard balls.

Moreover, the set of physically possible motions of the N real particles, on this theory, is going to be fantastically *smaller* than it is on the familiar Newtonian theory of $N+M$ real particles—and fantastically smaller than we intuitively *suppose* it to be—because the physically possible motions of those N real particles here are going to be restricted to the ones that are compatible with one particular set of initial conditions for the M fake ones.

And (finally) there is the very simple observation that it just *makes no physical sense* to distinguish, in the way we have been doing here, between the N "real" particles and the M "fake" ones—because the mathematical structure of the theory treats all of them in a completely identical way. And so (once again) the theory seems to be insisting that the M "fake" particles are in fact not one whit less real and concrete and substantial than the N "real" ones.

And let me remind you, yet again, that none of these sorts of worries arise in connection with the project of absorbing *space-time* into the laws, in the way that we were discussing above.

―――

Good. There's an analogous dialectic one can go through with (say) Maxwellian electrodynamics. This one turns out to be a little bit less one-sided, and a little bit more interesting, than the one about the unwanted particles. Suppose (then) that we like to think of *all* our particles—but not any of our *fields*—as real, concrete, physical things. Suppose (that is) that we should like to find some way of *absorbing* the fields into the laws of the motions of the *particles*. Here's a first attempt at doing something like that: **The physically possible motions of the particles are just the ones that have at least one faithful embedding into an Aristotelian space-time, decorated with an electromagnetic** field, such that the whole business (particles+fields) satisfies the equations of Maxwellian electrodynamics.[5]

This is manifestly going to have the same sorts of problems as the first of the strategies we considered above for getting rid of unwanted particles, which is that there really isn't anything much like a *dynamics* here *at all*—not a deterministic one and not a chancy one either. Suppose (that is) that there is some particular bump or dimple in the electromagnetic field—suppose (for example) that there is some particular *ray* of *light*—that never interacts with any of the particles until a certain time t_c—at which point it collides with real particle #14, altering particle 14's direction of motion. In that case, nothing in the motions of any of the particles, all the way from $t = -\infty$ to t_c, is going to offer any hint at all of what that ray of light is doing, or when or how or whether it might be expected to interfere with the motions of any of the particles. And so there is clearly going to be a continuous infinity of *different* trajectories of the particles, all of which are fully in accord with the laws of this theory, and all of which *exactly and entirely coincide with one another* for all times prior to $t = t_c$.

And if (in order to cure this disease) some *particular* initial electromagnetic field configuration is incorporated into the explicit statement of the law, then the law will be deterministic—but it will also be perversely *complicated*—much more so (indeed) than in the previous example with the particles. And here, again,

---

[5] There have been completely different kinds of attempts at writing down a version of classical electrodynamics without electromagnetic fields—attempts (that is) that have nothing to do with absorbing the fields into the domain of the laws. See, for example, Dustin Lazarovici, "Against Fields," *European Journal for Philosophy of Science* 8, no. 2 (May 2018): 145–170, for a recent and spirited defense of the "direct interaction" picture of Feynman and Wheeler.

there is going to be a vastly smaller set of physically possible evolutions of the particles than we are naturally inclined to think.

In this case—unlike in the case of the particle theory—there is not going to be a worry about our having chosen to *remove* something from the ontology that has exactly the same sort of mathematical representation in the theory as something *else* that we've chosen to *leave in*. But there is a more general worry that has much the same *form*, and that seems to me to have much the same *force*, to the effect that we are removing something here (that is: the field) that bears all of the familiar and unmistakable signatures of concrete physical *stuff*. I'm not sure how to say this with even remotely as much precision and explicitness as it needs to be said, but let me see if I can make a start.

There are two things that immediately pop into my head in this connection. One is an intuition to the effect that all of the chaos and ugliness and arbitrariness and complexity of the world has to do, almost as a matter of *conceptual analysis*, with the arrangement of the concrete physical *stuff*. The intuition is that it is of the very *essence* of stuff to be, in general, a mess—and it is of the very essence of *laws* to be clean and simple. And the other is that the fields that we are talking about eliminating here are the sorts of things that we are used to thinking we know how to *measure*. I will have a bit more to say about this last point in a minute.

―――⁓―――

Anyway, these questions of what counts as a physical *law* and what counts as a physical *thing* have become particularly serious and particularly urgent, of late, in the context of discussions of the ontology of the quantum-mechanical *wave-function*.

Let's think of wave-functions here—just so as to have some particular mathematical formalism explicitly on the table—in the context of Bohmian mechanics. Bohmian-mechanical wave-functions, somewhat like Maxwellian electromagnetic fields, push material particles around. But Bohmian-mechanical wave-functions, *unlike* Maxwellian electromagnetic fields, do not live in the three-dimensional space of our everyday experience of the world—*they* live, instead, in a space of $3N$ dimensions, where N is the total number of elementary particles in the universe.

And the business of incorporating $3N$-dimensional objects into the ontology of concrete physical things has made many investigators of the foundations of quantum mechanics distinctly and conspicuously uncomfortable. And there is a well-known and widely discussed strategy for *ameliorating* that discomfort—the strategy of so-called primitive ontology—which is to *remove* those objects from the category of concrete physical things, and to *absorb* them, in one way or another, into the category of *laws*.[6]

---

[6] Maybe it ought to be mentioned, at this juncture, that there is at least one *other* strategy for ameliorating this same discomfort available on the market nowadays—the so-called multi-field strategy. See, for example, M. Hubert and D. Romano, "The Wave-Function as a Multi-Field," *European Journal for Philosophy of Science* 8, no. 3 (October 2018): 521–537, and references therein.

Here's the idea: The wave-function is a concrete physical thing—the wave-function is (in particular) a concrete physical *field*—but unlike (say) the electric and magnetic fields of Maxwell, the wave-function takes on values not at individual *points* in three-dimensional space but (instead) at ordered *N-tuples* of points in three-dimensional space. In this way (so the thinking goes) we can have our cake and eat it too—we can hang on (that is) to the claim that the three-dimensional space of our everyday experience is the fundamental physical space of the world, *without* having to go to the trouble of moving the wave-function from the category of concrete physical *things* (where it intuitively seems to belong) to the category of abstract physical *laws*. Hubert and Romano (for example) are constantly saying things like: "In the [Bohmian-mechanical version of the] multi-field view, the scientific image consists of many particles in three dimensions, guided by a nonlocal field in this very space."

And the question is exactly what this last "in" is supposed to amount to. Surely the wave-function is not "in" the three-dimensional space in quite

And the obvious options for *doing* something like that are very much the same in the case of quantum-mechanical wave-functions as they were in the case of the electromagnetic field. We can write down laws of the motions of the Bohmian-mechanical particles that include an explicit specification of an initial wave-function, and we can write down laws of the motions of those particles that do *not* include an explicit specification of an initial wave-function. If the laws do *not* include an explicit specification of an initial wave-function, then (as in the previous examples we considered) there won't really turn out to be any dynamical laws *at all*. And if the laws *do* include an explicit specification of an initial wave-function, then they run the risk of being too complicated and too messy and too contingent-looking to be taken seriously. And again—as we found in both the case of our attempts to remove some of the particles from Newtonian mechanics, and as we found in the case of our

---

the same way as classical particles or electromagnetic fields are. It will make no sense (for example) to ask where within that space the wave-function is *located* (as one can do for a particle), or to ask where within that space the amplitude of the wave-function is large (as one can do for a Maxwellian electromagnetic field).

Maybe it would be better to say something like this: Even if the multi-field is not in any straightforward and familiar sense "in" the three-dimensional space—it is nevertheless a creature entirely *of* the three-dimensional space, in the sense that its *arguments* are built entirely out of *points* in that space.

But that's not right either. What's being overlooked, on this way of putting things, is that the argument of the multi-field is not a *simple* N-tuple of points in the three-dimensional space, but an *ordered* N-tuple of points in the three-dimensional space—and that the difference between two differently ordered N-tuples of the same N points in the three-dimensional space, the difference (for example) between $\{\alpha, \beta, \chi, \ldots \delta\}$ and $\{\beta, \delta, \chi, \ldots \alpha\}$, is not the sort of difference that is susceptible of being spelled out in terms of any intrinsic features of the three-dimensional space *itself*. The difference between $\{\alpha, \beta, \chi, \ldots \delta\}$ and $\{\beta, \delta, \chi, \ldots \alpha\}$ is (you might say) not a *three*-dimensional difference *at all*, but a $3N$-dimensional difference—and the natural *habitation* of a multi-field, no matter how you try to pretty it up, is apparently not a three-dimensional space, but a $3N$-dimensional one.

attempts to remove the fields from Maxwellian electrodynamics—the number and variety of physically possible evolutions of the Bohmian corpuscles is going to be much smaller, if we go with this deterministic route, than we would intuitively have thought.

~~~~~

Let's try something else—something that comes out of the literature on the foundations of statistical mechanics.

It is now more and more widely acknowledged, in the literature of the foundations of statistical mechanics, that the fundamental laws of nature include probability distributions over initial conditions. And this suggests a new and more sophisticated strategy for absorbing unwanted pieces of ontology into the laws—a strategy in which the laws into which the unwanted pieces of the ontology get absorbed turn out to be *stochastic*.

Suppose (for example) that there are M particles that we would like to remove from the universal catalogue of concrete material things by absorbing them into the laws. Start with the prohibitively complicated deterministic theory we discussed above—the one in which some particular set of initial conditions of the M particles that we want to remove are explicitly incorporated into the laws of the motions of the N particles that we want to hang on to. And consider a *probability distribution* over such deterministic "laws of motion"—the one induced by the usual statistical-mechanical probability distribution over initial conditions of the original $N+M$-particle universe—the one induced (that is) by the usual statistical-mechanical *Mentaculus* of the original $N+M$-particle universe.[7] This will yield a single *sto-*

---

[7] The one (that is) that you get by conditionalizing the standard statistical-mechanical distribution on the initial positions and velocities of the N "real" particles.

chastic law of motion for the remaining N "real" particles. Starting from any particular set of initial positions and velocities of the N "real" particles, this law will give us a probability distribution over future classical trajectories of those particles.

This escapes the obvious drawback of the deterministic strategy that we considered above—which is to say: it is *not* particularly complicated, and yet it manages to deliver us a real, dynamical, *law* of *motion*.

But this strategy gives us a theory that's bad in a different way. It gives us a theory that is in a curious kind of tension, at almost every turn, with *itself*.

Here's what I have in mind: On the one hand, this theory stipulates that M of its particles are *fake*—which is to say that it stipulates (among other things) that there are no determinate *facts* of the matter, at any particular time, about where those M fake particles *are,* or about how they are *moving*. But the theory also *entails* that these M fake particles are exactly as *detectable* as the N *real* ones—and it entails that the "positions" and the "velocities" of the M fake particles, at any particular time, can be *measured,* by exactly the same procedure, and with exactly the same mechanical instruments, as one uses to measure the positions and the velocities of the N real ones. Indeed—the theory entails that it is physically impossible to construct an instrument that is capable of *discriminating* between the real particles and the fake ones.

And it follows (for example) that nothing is going to stand in the way or our confirming, by experiment, and with every bit as much accuracy and with every bit as much certainty as we are able to do in the context of ordinary Newtonian mechanics, that the "measurable" properties of any isolated collection of real particles and fake particles, at any time t, are going to uniquely determine the positions of all of the real particles in that collection—and (while we're at it) of all of the fake ones too—at any *other* time

t', so long as the collection remains isolated between t and t'. And so there turns out to be a kind of *bad faith* about the formal "chanciness" of a theory like this. It is perfectly true, on this theory, that the histories of all of the officially concrete physical *items* in the world up to any particular time t do not uniquely determine the histories of those items at times *after t*— but it is *also* true that what you might call the *physically measurable properties* of the world, at any one time, on this theory, *do* uniquely determine the physically measurable properties of the world at all *other* times. And nothing like that is the case of any theory that we are ordinarily inclined to regard as *chancy*. And none of those paradigmatically chancy theories offers us any similarly obvious collection of candidates for reification that—if they were all to be reified—would give us something deterministic. Think (for example) of the GRW theory of the spontaneous collapse of the quantum-mechanical wave-function.

And the same thing happens with invariance under *time-translation*. The law of the motions of the N real particles, on a theory like this one, is going to have to include a probability distribution over the positions and velocities of the M fake particles that is stipulated to obtain at some particular *time*. And so, even if we are given a complete specification of the positions and the velocities and the internal properties of the N real particles in the world up to and including some particular temporal instant t, the probability that the theory associates with the proposition that the particles are going to swerve this way or that way just *after t*—as a result (for example) of colliding with one or another of the "fake" particles—is going to depend on what time t happens to *be*. And so there is an obvious official sense in which a theory like this is not invariant under time-translation. But the reader should by now have no trouble convincing herself that the world that this theory describes is nevertheless going to feature simple and universal and deterministic connections between the results of appropriately complete sets of measurements carried out at t and the results of appropriately complete sets of measurements carried out at t', which depend *only* on the value of $t'-t$, and *not at all* on the values of t or t' separately.

And this will be worth rubbing in some. The issue of time-translation invariance (for example) has no particular importance *in and of itself*. I myself have argued—elsewhere, and often—that the fundamental physical laws of a world like ours are going to need to make reference to a particular temporal moment. The point is not that there is anything intrinsically necessary or intrinsically valuable about time-translation invariance *per se*—the point (again) is just that the sort of theory we are considering here is somehow *at odds with itself* on that question. The point is that the connections that this theory imposes between the measurable properties of any isolated collection of real particles and fake particles, at two different times, are going to depend, not on *what two times those are*—but only on the temporal distance *between* them. And so there turns out to be something *disingenuous* about the official time-*dependence* of the dynamical laws of a theory like this.

And so there remains, even in this more sophisticated theory, a sense in which the mathematical structure of nature is palpably screaming at us that the "fake" particles—which have been absorbed into the law of the motions of the "real" ones—are really *concrete physical things:* And there is a unique and obvious and minimal *modification* of this theory that will eliminate the tension about the question of time-translation invariance, and there is a unique and obvious and minimal modification of this theory that will eliminate the tension about the question of determinism, and those two modifications turn out to be *exactly one and the same:* the *reification* of the M particles that have here been declared to be fake.

And we can play pretty much the same game with Maxwellian electromagnetism. We want to remove the electromagnetic fields from the category of concrete physical things—and we do so by absorbing them into the laws of the motions of the particles. We start (as above) with the prohibitively complicated deterministic theory we discussed above—the one in which some particular set of initial conditions of the electromagnetic fields that we want to remove are explicitly incorporated into the laws of the motions of the particles. And we consider a *probability distribution* over such deterministic "laws of motion"—the one induced by the usual statistical-mechanical probability distribution over initial conditions of all of the particles and electromagnetic fields in the universe. And this distribution will yield a single, simple, *stochastic* law of the motions of the particles—a law that (moreover) is not invariant under time-translations. But all of this turns out to be exactly as hollow in the case of Maxwellian electrodynamics as it was in the case on Newtonian mechanics:

The theory stipulates that there are no facts of the matter about the strengths and the directions of the electromagnetic fields at any particular point in space-time—but it also entails that those "strengths" and those "directions" can all be *measured* with instruments constructed entirely out of the *particles*. And so, as before, there is (to be sure) a literal and official sense in which the theory is stochastic, and there is (to be sure) a literal and official sense in which it fails to be invariant under time-translations—but it will also present every one of what you might call the *observable signatures* of being *precisely the opposite* of both of those things.

And things go pretty much the same way—of which more in a minute—when we play this game with Bohmian mechanics.

~~~

I have thus far alluded to three different proposals for absorbing Bohmian-mechanical wave-functions into a law of the motions of the Bohmian particles:

1) Stipulate that a given set of particle motions $\{x_1(t), x_2(t), \ldots x_N(t)\}$ is physically possible if and only if there is at least one solution to the Schrödinger equation and one initial configuration of the particles that—taken together with the Bohmian guidance condition— produces $\{x_1(t), x_2(t), \ldots x_N(t)\}$. On this proposal, the "law" of the motions of the particles turns out to be neither deterministic nor chancy—which is to say that it turns out not to be any familiar sort of law at all.

2) Consider some *particular* solution to the Schrödinger equation—call it $\Psi_{LAW}(X_1, X_2, \ldots X_N, t)$. That solution, together with the Bohmian guidance condition and the standard Bohmian rule connecting the probabilities of initial particle-configurations and the absolute squares of initial universal wave-functions, will give us a probability distribution $\rho_{LAW}(x_1(t), x_2(t), \ldots x_N(t))$ over all mathematically definable N-particle trajectories—and we now stipulate that *that* probability distribution is the complete law of the motions of the particles. On this proposal, the law of the motions of the particles is thoroughly *deterministic*—because any particular initial configuration of the particles $\{x_1(t=0), x_2(t=0), \ldots x_N(t=0)\}$, together with $\Psi_{LAW}(X_1, X_2, \ldots X_N, t)$ and the Bohmian guidance condition, will pick out a *unique* N-particle trajectory $\{x_1(t), x_2(t), \ldots x(t)_N\}$—but it is also incredibly *complicated*, and it will entail that the set of physically possible trajectories is much, much smaller than we ordinarily suppose.

3) Let $PH(\Psi_i(x_1, x_2, \ldots x_N, t))$ represent the standard statistical-mechanical probability distribution over the set of solutions to the Schrödinger equation $\{\Psi_1(x_1, x_2, \ldots x_N, t), \Psi_2(x_1, x_2, \ldots x_N, t), \ldots\}$ that are compatible with the Past Hypothesis. And let $\rho_i(x_1(t), x_2(t), \ldots x_N(t))$

represent the probability distribution over all mathematically definable N-particle trajectories induced by $\Psi_i(x_1, x_2, \ldots x_N, t)$ and the Bohmian guidance condition and the standard Bohmian rule connecting the probabilities of initial particle-configurations and the absolute squares of initial universal wave-functions (as in proposal 2). And now stipulate that the overall probability of the occurrence of any particular mathematically definable N-particle trajectory $(x_1(t), x_2(t), \ldots x_N(t))$ is equal to $\Sigma_i[\mathrm{PH}(\Psi_i(x_1, x_2, \ldots x_N, t))\, \rho_i(x_1(t), x_2(t), \ldots x_N(t))]$. On *this* proposal, the fundamental law of the motions of the particles is appealingly *simple,* but it is also *stochastic*—because the functions $\rho_i(x_1(t), x_2(t), \ldots x_N(t))$ and $\rho_j(x_1(t), x_2(t), \ldots x_N(t))$, with $i \neq j$, can both assign nonzero probabilities to N-particle trajectories that coincide before some time $t = \tau$ and diverge thereafter. This is the proposal I alluded to in the previous paragraph.

And I want to finish up by briefly discussing one further such proposal—the so-called *Wentaculus* picture proposed by my friend Eddy Chen.[8]

Eddy starts out with the same weighted combination of wave-functions—the $\mathrm{PH}(\Psi_i(x_1, x_2, \ldots x_N, t))$—that we used in the third of the proposals above. But his innovation is to replace the *old* Bohmian guidance condition—the one (that is) that gets us from $\mathrm{PH}(\Psi_i(x_1, x_2, \ldots x_N, t))$ + an initial N-particle configuration to a *probability distribution* over N-particle trajectories—with a *new* one that gets us from $\mathrm{PH}(\Psi_i(x_1, x_2, \ldots x_N, t))$ + an

---

[8] See, for example, Eddy Keming Chen, "The Past Hypothesis and the Nature of Physical Laws," in *The Probability Map of the Universe: Essays on David Albert's* Time and Chance, ed. Barry Loewer, Brad Weslake, and Eric Winsberg (Cambridge, MA: Harvard University Press, 2023), 204–248.

initial *N*-particle configuration to a *single* and *unique* and *determinate N*-particle trajectory, a trajectory that is essentially the *weighted average* of all of the trajectories that start out from that initial *N*-particle configuration in proposal 3.

So the laws of the motions of the particles are going to be every bit as *simple* on Eddy's picture as they are on proposal 3—but on Eddy's picture they are also going to be fully *deterministic*. What probabilities there *are* can all be traced back, as in ordinary Bohmian mechanics, to the familiar Bohmian probability distribution over the possible initial configurations of the *particles*. And Eddy's new Bohmian guidance condition is cooked up in such a way as to straightforwardly guarantee that the overall probability of the *N* particles in the world being configured in any particular way at any particular time is going to be exactly the same, on Eddy's theory, as it is on proposal 3.

Eddy's picture is certainly the most sophisticated attempt we have, as yet, at absorbing the Bohmian-mechanical wave-function into the category of the laws. But it should be noted, as well, that the diseases of our previous attempts in this direction have not yet been entirely expunged.

The simple and deterministic law of the motions of the particles—which is to say: the simple and deterministic law of the evolution of the entirety of the physical *world*, as far as Eddy's picture is concerned—is not invariant under time-translations. And the reason that this is something to worry about is not (again) because there is anything intrinsically *necessary* or intrinsically *valuable* about time-translation invariance *per se*—but (instead) because the question of invariance under time-translations puts the Wentaculus awkwardly at odds with *itself*. There is (on the one hand) a literal and official sense in which the laws of the motion the entirety of the physical world, according to the Wentaculus, explicitly depend on *what time it is*—and yet the world that this theory describes is going to feature simple and universal connections—*stochastic* connections, in this case—between the results of appropriately complete sets of measurements carried out at t and the results of appropriately complete sets of measurements carried out at t', that depend *only* on the value of $t'-t$, and *not at all* on the values of t or t' separately. And (once again) the instant the wave-function of the world is admitted back into the category of concrete physical things, everything magically snaps back into place: the laws of the motion of the world become formally and technically and genuinely invariant under time-translation, and the tension we have just now been describing disappears.

Let's see where this leaves us.

If one takes the *obvious* and *straightforward* and *flatfooted* approach to the business of interpreting one or another of the finished mathematical formalisms of quantum mechanics—if one approaches the business of interpreting one or another of the finished mathematical formalisms of quantum mechanics (that is) by asking: "What is this mathematical formalism *about*?" and "What is to this mathematical formalism as *particles* are to the formalism of *Newtonian mechanics*?" and "What is to this mathematical formalism as *charges* and *fields* are to the formalism of Maxwellian electrodynamics?" and "What are the *equations of motion* of this formalism the equations of the motion *of*?"— then the whole thing is pretty cut and dried, and the answer is manifestly going to consist, at least in part, of *the quantum-mechanical wave-function*.[9]

---

[9] The question of what *else* it may include is, of course, going to depend on what particular finished mathematical formalism we are talking about. In the case of the GRW theory, and also in the case of the Everett interpretation, we are going to be talking about the wave-function *simpliciter*. But in the case of Bohmian mechanics we are going to be talking about the wave-

And every attempt to *escape* from these obvious and straightforward and flatfooted interpretations—every attempt (in particular) to move the wave-function out of the category of *concrete physical things* and into the category of *laws*—seems to put this, or that, or everything, somehow *out of joint*.[10] And it goes without saying that this observation fits together nicely with the material in Chapter 1.

But there is no point in pretending that any of this, or all of it together, turns out to be *decisive*—and my purpose here was not so much to defend a position as to continue, and to deepen, and to expand, an existing conversation. The long and the short of it is that we are pretty badly in need—especially in the foundations of quantum mechanics—of some better and deeper and more stable and more deliberate understanding than is presently available of what sorts of things fit naturally into the category of the nomic, and what sorts of things fit naturally into the category of the concrete, and what sorts of things might be freely deposited, in the light, or under the pressure, of other sorts of considerations, into whichever of those categories we wish. And the question is not trivial.

---

function and the *particles*—and the "equations of motion" in question will be both the Schrödinger equation and the Bohmian guidance condition.

[10] We have, of course, been working here with the formalism of Bohmian mechanics—but the reader will have no trouble confirming for herself that the same conclusions will apply to (say) the "flash" version of the GRW theory, and to the "mass-density" version of the GRW theory, and to the "mass-density" version of the Everett interpretation, and so on.

3

The Still More Basic Question

Disclaimer: It seems only fair to warn the reader that there may not be anything going on here, at the end of the day, aside from my confessing to various deep philosophical confusions. I take the trouble of writing these confusions down because some of us who work on the foundations of physics have been painfully floundering around in them for something like a quarter of a century now, and because the business of getting them sorted out has come to look more and more essential to the business of understanding what it is that quantum mechanics is telling us about the world.

Think of it (then) as a cry for help.

The Question

Suppose that somebody were to propose that space is actually *ten*-dimensional. The proposal (in particular) is that there are seven more spatial dimensions built (as it were) *on top* of the three that we know about. The proposal is that the ten-dimensional space *contains* the familiar three-dimensional space of our everyday empirical experience as a proper geometrical *subspace*, and that the to-ings and fro-ings of the familiar material constituents of the world are *confined* to that subspace, and that they unfold *within* that subspace in exactly the way we are used to. So (for example) there is a perfectly straightforward sense in which the three-dimensional tables and chairs and rocks and trees of our everyday experience take up disjoint three-dimensional *regions* of that subspace, and there is a perfectly straightforward sense in which the proper physical *parts* of those tables and chairs and rocks and trees take up proper *geometrical* parts of those regions, and there is a perfectly straightforward sense in which that three-dimensional subspace as a whole consists of precisely that locus of points at which I might imaginably place (say) the tip of my finger. Maybe the theory says that there are *other* concrete physical things—things that we never see, and with which we have no causal connections—moving around in other, disjoint, subspaces of the ten-dimensional space. Or maybe it says that the ten-dimensional space is empty outside of our own particular three-dimensional slice of it.

It goes without saying that there would be no scientifically respectable reasons for *believing* a theory like that, but there would be no difficulty about the business of making *sense* of it—nobody is going to deny (that is) that it presents us with an intelligible and empirically adequate picture of the world. If the theory were to be put forward (for example) as a skeptical scenario, it would have to be acknowledged that there are no strictly logical or conceptual or empirical means of definitively *ruling it out*.

Now, the picture of the world that we end up with if we treat the quantum-mechanical wave-function as concrete fundamental physical *stuff* is high-dimensional too—but in this case the departure from the familiar three-dimensional manifest image of the world is much more radical. In the case of the nonrelativistic quantum mechanics of particles, for example, the points in the high-dimensional space in which a wave-function does its to-ing and fro-ing are going to correspond to *ordered N-tuples* of points in the three-dimensional space of our everyday experience—where N is the number of elementary particles in the world. So in *this* case, the three-dimensional space of our everyday experience is *not* going to be a subspace of the higher-dimensional one, and there is going to be *no sense at all* in which the tables and chairs and rocks and trees of our everyday experience take up disjoint three-dimensional *regions* of that higher-dimensional space, and there is going to be *no sense at all* in which *any* of the points in the higher-dimensional space are points at which I might imaginably place (say) the tip of my finger.

And all of the floundering around that I alluded to above has to do with questions of whether a fundamental physical theory that departs from the familiar three-dimensional manifest image of the world as radically as that is even susceptible of being *entertained*. All of the floundering around that I alluded to above (that is) has to do with questions of whether there is even so much as a conceptual or metaphysical *possibility* that a world that consists of fundamental stuff to-ing and fro-ing in a high-dimensional space like the one described in the previous paragraph can accommodate the rocks and chairs and tables and trees of our everyday experience. And those are the sorts of questions that I want to consider here.

Let's focus on a simple representative example. Here are two hypothetical fundamental physical theories:[1]

Theory 1 describes a world that consists of N familiar Newtonian particles, floating around in an infinite three-dimensional

---

[1] I am going to be discussing thirteen different hypothetical fundamental physical theories in this essay—many of which have similar names. The reader can find a complete glossary of those theories at the end of this essay—listed in the order in which they first appear.

Euclidian space, under the influence of a classical Hamiltonian of the form:[2]

$$H = \left\{ \sum_i^N (m_i((dx_i/dt)^2 + (dy_i/dt)^2 + (dz_i/dt)^2)/2) \right\}$$
$$+ \left\{ \sum_{k \neq j}^N V_{kj}((x_k - x_j)^2 + (y_k - y_j)^2 + (z_k - z_j)^2) \right\}. \quad (1)$$

And I want the reader to pretend that the world described by Theory 1 can accommodate the rocks and trees and tables and chairs and people and haircuts and baseball games and all of the paraphernalia of our everyday macroscopic experience. It can't, of course. That's why we need theories like quantum mechanics. But it will do no harm, and it will keep things helpfully simple, to pretend, for the moment, that it can.

Theory 2 describes a world that consists of a single point-like physical item—call it (after Shelly Goldstein) the *marvelous point*—floating around in an infinite $3N$-dimensional Euclidian space, under the influence of a classical Hamiltonian of the form:

$$H = \left\{ \sum_i^{3N} (m_i((dx_i/dt)^2/2)) \right\}$$
$$+ \left\{ \sum_{k \neq j}^N V_{kj}((x_{3k} - x_{3j})^2 + (x_{3k-1} - x_{3j-1})^2 + (x_{3k-2} - x_{3j-2})^2) \right\} (2)$$

where $x_1 \ldots x_{3N}$ is some Cartesian coordinatization of the $3N$-dimensional space, and $m_{3i} = m_{3i-1} = m_{3i-2} = m_i$ for all $i = 1 \ldots N$, and where $V_{kj}((x_{3k} - x_{3j})^2 + (x_{3k-1} - x_{3j-1})^2 + (x_{3k-2} - x_{3j-2})^2)$ is the same function of the arguments $((x_{3k} - x_{3j})^2 + (x_{3k-1} - x_{3j-1})^2 + (x_{3k-2} - x_{3j-2})^2)$ as $V_{kj}((x_k - x_j)^2 + (y_k - y_j)^2 + (z_k - z_j)^2)$ is of the arguments $((x_k - x_j)^2 + (y_k - y_j)^2 + (z_k - z_j)^2)$.

The worlds that these two theories describe are, on the face of it, very different. In the world of Theory 1, the fundamental physical objects are particles, floating around in a three-dimensional

---

[2] See footnote 1 of the first essay in this volume.

space, whereas the world of Theory 2 consists of a *single* fundamental physical object—the "marvelous point"—floating around in a $3N$-dimensional space. But there is (for all that) an obvious and exact and complete formal *congruence* between them. There is (to begin with) a pretty simple one-to-one mapping between the possible configurations of the N particles in the three-dimensional space and the possible locations of the marvelous point in the $3N$-dimensional space (where the first three coordinates of the marvelous point correspond to the x, y, and z coordinates, respectively, of particle 1, and the second three coordinates of the marvelous point correspond to the x, y, and z coordinates, respectively, of particle 2, and so on). And that mapping will induce (in its turn) a one-to-one mapping between the *solutions* to the *equations of motion* of Theory 1 and the solutions to the equations of motion of Theory 2.

And the question I want to consider is what to *make* of that mathematical resemblance. The question I want to consider is (more particularly) whether or not, in the *light* of that mathematical resemblance, and given the stipulation that Theory 1 can accommodate rocks and trees and tables and chairs and people and haircuts and so on, Theory 2 can accommodate all those things as well.

Puritanism

Some say not. They say that there are no rocks or trees or tables or chairs, or anything of the kind, in any of the worlds described by Theory 2. They say that the simple one-to-one mapping from the solutions of Theory 2 to the solutions of Theory 1 is a merely *mathematical* resemblance, a merely *structural* resemblance, from which nothing in the way *physical* or *metaphysical* resemblances can be inferred.

Call them the puritans.

According to the puritans, asking "What is it that a marvelous point would need to *do,* what kind of a *trajectory* would it need to *follow,* all alone in a $3N$-dimensional space, in order to for there to be rocks and trees and tables and chairs and people?" is like asking "How many apples would it take for there to be an orange?" The thought is that apples simply have nothing to *do* with oranges. The thought is that there is manifestly *no* number of apples, to-ing and fro-ing in any way you like, that are ever going to amount to an orange. And there is an obvious and analogous and unbridgeable explanatory and metaphysical gap, so the puritans say, between a lone marvelous point, floating around in a $3N$-dimensional space, and anything in the way of tables and chairs and rocks and trees and people.[3]

The puritans are very much impressed with the fact that once the raw fundamental objects of Theory 1 are merely, visually, *set before you,* all you need to do to make out the tables and chairs and buildings and people is (as it were) to *squint,* or to *step back,* or to *coarse-grain*[4]—and they are suspicious of any conceptions of the relationship between the fundamental and the nonfundamental that are fancier or more abstract or less primordial than that.

But that can't be right. What about (say) *heat?* No amount of *squinting* or *stepping back* or *averaging* or *coarse-graining* is ever

---

[3] Whether there are any perfectly realized puritans in the world is not (it ought to be acknowledged) altogether clear. Something in the general neighborhood of puritanism seems to me to be at the bottom of a lot of the talk, in the foundations of quantum mechanics, about "primitive ontology"—but the adherents of primitive ontology are not a monolithic group. If I had to guess, I would say that Shelly Goldstein might well be a perfect puritan, whereas Tim Maudlin is probably more of a puritan sympathizer.

[4] For a particularly vivid example, see fig. 1, and the text around it, in Tim Maudlin, "Can the World Be Only Wavefunction?," in *Many Worlds? Everett, Quantum Theory, and Reality,* ed. Simon Saunders et al. (Oxford: Oxford University Press, 2010), 123.

going to make *motion* look *hot*. The way we come to connect heat with motion is by means of a much more *abstract* and *circuitous* and *functional* kind of an analysis: We find that molecular motion is the thing that causes certain sorts of excitations in our sensory neurons, and we find that molecular motion is the thing that causes the readings of thermometers to rise, and we find that molecular motion is the thing that causes ice to melt, and (more generally) we find that molecular motion is the thing that plays the heat *role*, we find that molecular motion is the thing that occupies the heat *node*, in the vast universal network of causal relations. And this connection between heat and motion—even though it has nothing very literal to do with *seeing*—is (it goes without saying) one of the exemplary and paradigmatic and definitive triumphs of scientific *understanding*.

Global Functionalism

Let's try something else.

Consider the sort of functional analysis whereby we arrive at a reduction of heat to motion. Suppose we were to apply that sort of analysis to *everything*—or (rather) to everything that isn't *fundamental*. Suppose (that is) that we say that there is nothing more or less to being a rock, or a tree, or a chair, or a person, or a haircut, or a lawsuit, or a university, or a molecule, than to have a certain *causal profile*—to occupy a certain *node* in the overall network of causal relations. Call that *global functionalism*. The talk of "nodes" and "profiles" and "the overall network of causal relations" is no doubt a little vague, as it stands—but suppose that we are willing to stipulate that on any *sensible* conception of the "overall network of causal relations" of this or that possible physical world, the isomorphism between Theory 1 and Theory 2 is going to guarantee that those two theories share the same overall network of causal relations. Then the global functionalist idea of what it is to be a table or a chair or a rock or a tree is going to entail that if Theory 1 can accommodate the existence of the tables and chairs and rocks and trees of our everyday experience of the world, then, necessarily, and a priori, and as a matter of pure conceptual analysis, Theory 2 can accommodate all of that as well.

Let's see if we can sharpen this up into something a little more detailed and precise.

Let's confine ourselves—just to keep things simple, and just for the sake of this particular conversation—to fundamental physical theories that consist of a specification of a fundamental physical *space*, and of a fundamental physical ontology of structureless point-like physical items *floating around* in that space, and of fundamental *dynamical laws* that govern the *motions* of those items in that space. And note—and this is going to be crucial to what follows—that every theory like that is going to bring with it a space of what you might call possible *dynamical conditions*. The *dynamical condition* of the world, at any particular temporal instant, is a specification of all of the information about the world at the instant in question—or all of the information about the world that can in one way or another be *uniquely attached* to the instant in question (by way of, say, derivatives with respect to time)—that is required in order to bring the full predictive resources of the fundamental dynamical laws to bear. And the *space* of the possible dynamical conditions of the world, in the sorts of fundamental physical theories that we are considering here, is referred to in the scientific literature as the *phase space* of the theory in question.

A few examples will make all this terminology clear:

1) Theory 1 (T_1): A theory (that is) of N fundamental Newtonian particles, floating around in a three-dimensional fundamental physical space, under the

influence of a Hamiltonian like the one in equation (1). The *phase space* of T_1—the space (that is) of the possible *dynamical conditions* of the kind of world that T_1 describes—is a $6N$-dimensional space, each point of which corresponds to a complete specification of the $3N$ coordinates and the $3N$ *momentum-components* of the N fundamental particles.

2) Theory 2 (T_2): A theory (that is) of a single point-like physical item (the "marvelous point") floating around in a $3N$-dimensional fundamental physical space, under the influence of a Hamiltonian like the one in equation (2). The phase space of T_2—the space (that is) of the possible dynamical conditions that T_2 describes—is also a $6N$-dimensional space, each point of which corresponds to a specification of the $3N$ coordinates of the marvelous point and the $3N$ quantities $m_i(dx_i/dt)$.

3) A theory—call it Theory 3 (T_3)—of a single fundamental point-like physical item floating around in a $6N$-dimensional fundamental physical space. The dynamical laws of Theory 3 stipulate that the possible *trajectories* of this point-like item through this $6N$-dimensional fundamental physical space are precisely the possible trajectories of the *phase point* of Theory 1 through its $6N$-dimensional *phase space* (which, by the way, are *also* precisely the possible trajectories of the phase point of Theory 2 through *its* $6N$-dimensional phase space). And note that the *phase* space of Theory 3 is going to be identical to its *fundamental physical* space—since (in Theory 3) the position of the fundamental point-like physical item in the fundamental physical space at any given temporal instant is going to uniquely determine the *trajectory* of that item, through that space, from $t = -\infty$ to $t = \infty$.

So in some theories (like Theory 1 and Theory 2) the phase space is *different* from the fundamental physical space, and in others (like Theory 3) the phase space turns out to be *identical* to the fundamental physical space. And note (as well) that although these three theories describe distinct and incompatible worlds—with distinct and incompatible fundamental physical *spaces,* and distinct and incompatible fundamental physical *ontologies*—their *phase* spaces, and their respective sets of physically possible *trajectories through* those phase spaces, all turn out to be the same.

Good.

Consider two theories of the general sort that we have been talking about—theories (that is) of the motions of point-like fundamental physical items in some fundamental physical space—T_a and T_b. And let S_a be the phase space of T_a, and let S_b be the phase space of T_b. And let $^aU_\tau$ be the dynamical operator that carries the phase point of the world τ seconds forward in Theory T_a, and let $^bU_\tau$ be the dynamical operator that carries the phase point of the world τ seconds forward in Theory T_b. And suppose that there is some one-to-one mapping M_{ba} between the points in S_a and the points in S_b that satisfies:

1) For every point y in S_b, if $M_{ba}(y) = x$, then $M_{ba}(^bU_\tau(y)) = {^aU_\tau}(x)$, and
2) For every point x in S_a, if $M_{ba}(x) = y$, then $M_{ba}(^aU_\tau(x)) = {^bU_\tau}(y)$.

Whenever all of that is the case—whenever (that is) the relationship between T_a and T_b is as illustrated in Figure 3.1—we will say that T_a and T_b are *dynamically equivalent* to each other.

And it will also be convenient, while we're at it, to define a slightly more general sense in which the dynamics of one theory can "contain" the dynamics of another, which runs as follows:

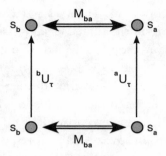

FIGURE 3.1

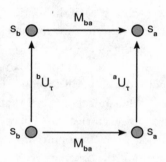

FIGURE 3.2

Whenever there is some mapping M_{ba} from some subset of the points in S_b to all of the points in S_a that satisfies

  i) For every point x in S_a, there is at least one point y in S_b such that $M_{ba}(y) = x$, and
  ii) For every point y in S_b, there is at most one point x in S_a such that $M_{ba}(y) = x$, and
  iii) For every point y in S_b, if $M_{ba}(y) = x$, then $M_{ba}(^bU_\tau(y)) = {}^aU_\tau(x)$,

we say that the dynamics of T_b *contains* the dynamics of T_a. That situation is illustrated in Figure 3.2. These notions of equivalence and containment are (of course) closely related to one another: T_a and T_b are dynamically equivalent to one another if and only if they contain each other's dynamics.

Now, the global functionalist idea is that what it is to be any particular nonfundamental thing—a table (say), or a chair, or a molecule, or a haircut—is nothing more or less than to occupy this or that *node* in the overall *causal network* of the universe. And I have already noted that this talk of "nodes" and "networks" is, as it stands, a little vague. But *whatever* all the talk of nodes and networks and profiles might turn out to mean, it seems innocent enough to suppose that if T_b contains the *exact and complete dynamics* of T_a—in the sense of Figure 3.2—then T_b must also contain the *overall causal network* of T_a. And so it is apparently going to follow from the global functionalist analysis of what it is to be a table or a chair or a molecule or a haircut that if T_a can accommodate tables and chairs and molecules and haircuts, and if T_b contains the dynamics of T_a, then T_b can accommodate tables and chairs and molecules and haircuts too.

For the global functionalist, a mapping that satisfies (i)–(iii), *any* mapping that satisfies (i)–(iii), is going to amount (you might say) to a *magnifying glass,* or to a cleverly curved *mirror,* or to a helpful way of *coloring things in,* by means of which one can see that the world of T_a was always already *there* in the world of T_b.

And note that Theory 1 and Theory 2 and Theory 3 are all, by the above definitions, and notwithstanding their very different fundamental spaces and ontologies, dynamically *equivalent* to one another. And it is going to follow—according to the global functionalist—that if Theory 1 can accommodate tables and chairs and molecules and haircuts, then, *necessarily* and *a priori,* Theory 2 can too.

But so can Theory 3.

And this way (I think) lies madness.

Consider a theory—call it Theory 4 (T_4)—of a single structureless point-like fundamental physical item in a Q-dimensional space. And suppose that the single fundamental dynamical *law* of this world is that the point-like item *never moves*. And note that—if Q happens to be divisible by 6—the position of an item like that could be treated as *encoding* the positions and momenta—at some "initial" time t_0—of $Q/6$ Newtonian particles floating around in a three-dimensional space in accord with the dynamical laws of T_1. And in this way one could set up a *mapping* $M_{41}(y, t) = x$—which will depend on time, and on the dynamical laws of T_1—that takes us from any point y in the phase space of Theory 4 to a corresponding point x in the phase space of Theory 1, and which straightforwardly satisfies (i)–(iii). And this mapping is clearly *invertible*. And so Theory 1 and Theory 4 turn out to be *dynamically equivalent* to one another.

This kind of mapping is (by the way) a beautiful and familiar tool of classical analytical mechanics, where it is known as a *Hamilton-Jacobi* transformation—and the analogue of such transformations in quantum mechanics is precisely the *Heisenberg picture,* in which the physical system (which is to say: the universe) remains permanently at rest in its Hilbert space, and it is (rather) *the connection between the operators and the observables* that evolves in time.

Anyway—what's relevant for our purposes here is that the mere mathematical *existence* of such a mapping, the mere *dynamical equivalence* (that is) of Theory 1 and Theory 4, is apparently going to commit the global functionalist to the claim that the world of Theory 1 is really, seriously, metaphysically, already *there*, in its entirety, in the world of Theory 4. What's relevant—to put it a little more vividly—is that the mere mathematical existence of such a mapping is apparently going to commit the global functionalist to the claim that *all that God had to do*, in order to bring into being all of the particles and rocks and trees and tables and chairs and people and lawsuits and baseball games and the very three-dimensional space of our everyday experience of the world, is to place a single structureless point-like physical item at rest in a sufficiently high-dimensional fundamental space, and to stipulate that the single law of the evolution of the world is that *nothing ever happens*.

Let's press this a little further. Consider a theory—call it Theory 5—of the motions of $Q/6$ Newtonian particles floating around in an infinite three-dimensional Euclidian space and having no interaction with one another *at all*. Or another one—Theory 6—which is a theory of the motions of $Q/6$ Newtonian particles floating around in an infinite three-dimensional space and interacting with one another only by way of very short-range mutually repulsive forces—like billiard balls. And note that we can set up a Hamilton-Jacobi mapping from Theory 4 to Theory 5—and another one from Theory 4 to Theory 6—in just the same way as we previously set up a mapping form Theory 4 to Theory 1 (that is: by reading the position of the item in Theory 4 as fixing the initial positions and velocities of the particles in Theory 5 or Theory 6).

And so the global functionalist is now apparently committed to the claim that if God were to place a single structureless point-like physical item at rest in a sufficiently high-dimensional fundamental space, and to stipulate that the single law of the evolution of the world is that *nothing ever happens*, she would thereby necessarily and unavoidably bring into being *not only* the tables and chairs and universities and haircuts of Theory 1—but also (who knows where) the universal featureless gases of Theories 5 and 6! And as a matter of fact—since dynamical equivalence is a transitive relation—the global functionalist is now going to be committed to the claim that God could not have made either of the universal gases of Theories 5 or 6 without *also* making all of the tables and chairs and universities and haircuts of Theory 1.

And all of this, I take it, is absurd.

Let's see if we can come up with a sensible way of reining it in.

Let's begin by reviewing how it is that we got in trouble. Suppose that Theory 1 can accommodate the tables and chairs and rocks and trees of our everyday experience of the world. And suppose (as the global functionalists do) that all there is to being a table or a chair or a rock or a tree is to "occupy" the relevant "node" in the "overall causal network of the world." And suppose that we agree that any smooth invertible mapping from the phase space of one classical-mechanical theory to the phase space of another classical-mechanical theory that satisfies requirements (i)–(iii) is going to preserve the "overall causal network of the world" (whatever, exactly, "the overall causal network of the world" may turn out to mean). Then it is going to follow that theories like Theory 4 and Theory 5 and Theory 6 can accommodate the tables and chairs and rocks and trees of our everyday experience as well.

Maybe the thing to do (then) is to tighten up the analysis of what it is to be a table or a chair or a rock or a tree. Maybe we ought to say (in particular) that there is more to being a table or a chair or a rock or a tree than merely occupying the appropriate node on the overall causal network of the world—maybe we ought to require (in addition) that the business of specifying what particular to-ings and fro-ings of the fundamental stuff of the world give us what particular arrangements of tables and chairs and rocks and trees not be too (I don't know) *complicated*, or too *grotesque*, or too *opaque*, or something like that.[5]

---

[5] David Wallace presumably has something like this in mind when he says that the mapping from the fundamental language to the language of tables and chairs and tigers and so forth should be "relatively simple." David Wallace,

One way to think of the puritans (for example) is as imposing a particularly extreme and particularly *literal* form of the prohibition on *opaqueness*. And the thought here is that maybe there is a sensible analysis of what it is to be a table or a chair or a rock or a tree that is more liberal than the analysis of the puritans but less liberal than the analysis of the global functionalists. This intermediate analysis would presumably endorse the familiar scientific understanding of *heat*, and it might also allow that Theory 2 can accommodate the tables and chairs and rocks and trees of our everyday experience of the world, but it would *deny* that theories like Theory 4 or Theory 5 or Theory 6 can accommodate things like that. Or maybe it would say that whatever rocks and trees and tables and chairs there are in theories like Theory 4 or Theory 5 or Theory 6 are somehow *second rate*, or that they are somehow *less real* than the ones we find in theories like Theory 1 or Theory 2.

The devil here, I guess, is going to be in the details. The business of being "not too complicated" seems much too vague and too arbitrary and too dull to serve as an instrument for distinguishing tables that are *real* from tables that aren't. And if the "simplicity" in question here is supposed to have anything to do with usefulness in the practical everyday business of doing physics—then *all* of the mappings alluded to above—not just the one from Theory 2 to Theory 1, but the ones from Theory 4 and Theory 5 and Theory 6 to Theory 1 as well—are the very *model* and *paradigm* of simplicity. And I don't think I understand what it would even *mean* to say of some pair of tables that they are real to different *degrees*.

Here's another thought: Maybe there ought to be a prohibition on any explicit *time-dependence* in the mapping from the

---

The Emergent Multiverse: Quantum Theory according to the Everett Interpretation (Oxford: Oxford University Press, 2012), 54.

fundamental vocabulary to the nonfundamental vocabulary. This would at least have the virtue of being sharp and clear and explicit (in contrast with the prohibitions on "complexity" and "grotesqueness" and "opacity" that we were talking about before). And it seems to have a principled and intuitively compelling motivation—that the world changes only when there are changes in the arrangement of the totality of the *fundamental stuff*.[6] And this will endorse the standard scientific analysis of heat, and it will allow that Theories 2 and 3 can accommodate the tables and chairs and rocks and trees of our everyday experience of the world—but it will decisively *block* the argument that Theory 4 can accommodate those things as well.

This apparent success (however) turns out to be very narrow, and very short-lived.

Consider (for example) the following modification of Theory 4—call it Theory 4*: In Theory 4*, the structureless point-like physical item is sitting not in a Q-dimensional space but in a $(Q+1)$-dimensional space. And the additional dimension is called f. And the single fundamental dynamical *law* of the world—in Theory 4*—is not (as above) that *nothing ever happens* but that the point-like item moves, forever, with a velocity of exactly 1, in the f-direction.

In Theory 4*, the position of the item along the f-axis can serve as a *clock*—and so the time-dependent mappings $M_{41}(y,t)=x$ and $M_{45}(y,t)=x$ and $M_{46}(y,t)=x$ and what have you can all be swapped out for $M_{4*1}(y,f)=x$ and $M_{4*5}(y,f)=x$ and $M_{4*6}(y,f)=x$, and what have you, respectively, which are all, of course, time-*independent*. And so it will follow (for example)—even by the lights of the more restrictive sort of global functionalist that we are considering here—that it is not within the power of God to create the world of Theory 4* without, at the same time, and by

---

[6] This is the sort of thing I myself used to think.

the same token, and in the same act, also creating the world of Theory 1 and the world of Theory 2 and the world of Theory 3 and the world of Theory 5 and the world of Theory 6 and an unmanageable infinity of other very different-looking worlds as well. And so all of the absurdity remains (behold!) intact.[7]

And those are all of the suggestions along these lines that I can think of, or have heard about.

Scientific Functionalism

So the functionalist argument that Theory 2 has particles in it seems to commit us to all sorts of absurdities. Does the same apply to the functionalist argument that heat is a certain kind of

---

[7] Or (rather) *almost* all of it does. Note (in particular) that Theory 4* is not dynamically *equivalent* to Theories 1 and 2 and 3 and 5 and 6—as Theory 4 was. Theory 4* (instead) *contains* the dynamics of Theories 1 and 2 and 3 and 5 and 6. And although that containment suffices to guarantee that God could not create the world of Theory 4* without also creating the worlds of Theories 1 and 2 and 3 and 5 and 6—it will not suffice to guarantee that God could not create the world of Theory 5 without also creating the worlds of Theories 1 and 2 and 3 and 4* and 6. But even that can be fixed. There is (it turns out) a fully invertible and fully time-independent Hamilton-Jacobi transformation that takes us from a theory of a single structureless point-like physical item that moves, forever, with velocity 1, along a straight line, in a $6N$-dimensional space—call it Theory Super-4— to a theory of N Newtonian particles, moving around in a three-dimensional space, under the influence of any classical Hamiltonian you like. A particularly beautiful example of a transformation like this is presented in Cornelius Lanczos, *The Variational Principles of Mechanics* (Toronto: University of Toronto Press, 1949), 231–233. And the mere mathematical existence of a transformation like *that* is going to suffice—even by the lights of the more restrictive sort of global functionalist that we are considering here—to show that Theories 1 and 2 and 3 and 5 and 6 and Super-4 are all dynamically *equivalent* to one another. And that dynamical equivalence will *in turn* suffice—even by the lights of the more restrictive sort of global functionalist that we are considering here—to show that God could not have created the world of any *one* of those theories without also, at the same time, and by the same token, and in the same act, creating the worlds of all of the others.

motion? I don't think so. There is (I think) an important difference between the two arguments. And it will be worth taking a minute to see how that plays out.

The argument that Theory 2 has particles in it proceeds, as we have seen, as follows: One notes (to begin with) that the phase space of Theory 1 is *identical* to the phase space of Theory 2. And so there is a trivial a one-to-one mapping from the solutions to the equations of motion of Theory 1 to the solutions to the equations of motion of Theory 2. And it follows that Theory 1 and Theory 2 are *dynamically equivalent* to each other. And it follows that the "overall causal network" of Theory 1—whatever, exactly, that might turn out to be—is identical to the overall causal network of Theory 2. And it follows that if Theory 1 can accommodate particles and tables and chairs and rocks and haircuts, and if all it is to be a particle or a table or a chair or a rock or a haircut is to occupy a certain "node" in the overall causal network of the universe, then the universe of Theory 2 must be able to accommodate all of that stuff as well.

This argument is meant to show us how to descry the particles and tables and rocks and what have you of Theory 1 in the to-ings and fro-ings of the marvelous point in Theory 2 by showing us how to descry, at a single stroke, the *entirety of the universe* of Theory 1 in the to-ings and fro-ings of the marvelous point of Theory 2. The descrying is not supposed to work *piece by piece* (that is) but *all at once*. The descrying of the tables depends on the descrying of the chairs, and the descrying of the chairs depends on the descrying of the tables, and the descrying of the tables and the chairs depends on the descrying of the particles, and the descrying of the particles depends on the descrying of the tables and the chairs. We descry the tables and the chairs and the trees and the lawsuits and the particles not individually, and not in any particular order of logical dependence, but as a single corporate body. The descrying is *global*, and (you might say) *unanchored*.

And the *worry* about this sort of argument is (again) that the business of bringing the causal web of this or that fundamental physical theory into this kind of correspondence with the causal web of Theory 1 turns out to be much too *easy*. The worry (that is) is that if we insist that there is nothing more to being a table than occupying a certain node in a certain abstract universal causal structure, then we are going to end up with tables, and with rocks and trees and buildings and particles and haircuts and lawsuits and temperatures and colors and chemical properties as well, almost anywhere you look—and (in particular) in Theories 2 and 3 and 4 and 5 and 6.

And the kind of functional analysis that we turn to again and again in the course of everyday *scientific practice*—the kind of functional analysis whereby we discover (for example) that *temperature* is a measure of *speed*—works in a very different way. The important kernel of truth in the functional style of analysis—if I understand it correctly—is the observation that we *pick things out*, that we *distinguish things from one another*, in terms of their different physical effects on *us*. The important grain of truth in the functional style of analysis—to put it a little more generally—is the observation that we distinguish things from one another in terms of their different physical effects on a certain set of what you might call *anchors* or *reference objects* (the external surfaces of our bodies, our measuring instruments, our epistemic community, various medium-sized dry goods, what have you), whose privileged status, at least for the sake of this or that particular conversation, is treated as something antecedently and unproblematically *given*.

What we say about *temperature*, in everyday scientific practice, is not that it is that thing that occupies a certain node in the abstract and uninterpreted universal causal network of the world, but (instead) that it is that thing that stands in a particular set of causal relations to things like nerve endings and thermometers and so on.

On *this* kind of functional analysis (then), what counts as heat depends on what counts as a nerve ending, but not the other way around. This kind of functional analysis extends itself outward into the world by means of a succession of *local* steps, rather than a single *global* one. This kind of functional analysis (to put it slightly differently) is *anchored* rather than *unanchored*—and it features a very definite *order* of *logical dependency*.

And note that this *local* kind of functional analysis—unlike the *global* one—will make it possible to *endorse* claims like "Heat is a form of motion," or claims like "Theory 2 has tables and chairs and particles in it" without *also* endorsing claims like "Theory 4 has tables and chairs and particles in it." What we endorse and what we don't—in these sorts of contexts—is going to be a matter of which translations from the language of the fundamental to the language of the nonfundamental we are willing to treat as somehow unproblematically *given*—for the sake, or in the context, of whatever conversation it is that we happen to be involved in. And note that just the tiniest bit of anchoring can sometimes yank a whole world of medium-sized dry goods into being: If we are willing to treat it as somehow unproblematically given (for example) that a certain single triplet of the degrees of freedom of Theory 2—(x_9, x_8, x_7), say—are the three degrees of freedom of a Newtonian particle, then it will follow from the *local* and *commonsensical* and *scientific* variety of functional analysis that every *other* triplet of the form (x_k, x_{k-1}, x_{k-2}) will be the three degrees of freedom of some *other* Newtonian particle, and that there will be tables and chairs people and haircuts and universities to boot. But nothing whatever about the existence or nonexistence of particles or tables or chairs or what have you is going to follow—on *this* kind of analysis—from the causal structure of Theory 2 *in and of itself.*

So scientific functionalism tells us a great deal, in what you might call a *conditional* way, in what you might call a *contex-*

tual way, about *what it is* to be a particle or a rock or a tree or a haircut. But it is apparently not going to get us any closer to settling the question that we are primarily wondering about here—which is not the question of whether or not a fundamental theory like Theory 2 can accommodate rocks and trees and haircuts *given* that it can accommodate (say) *particles and universities and toothbrushes*—but the question of whether or not a fundamental theory like Theory 2 can accommodate particles and rocks and trees and haircuts and universities and toothbrushes, and so on, *simpliciter*.[8]

Libertinism

The global functionalists were hoping to find an a priori conceptual analysis of what it is to be a table or a chair or a rock or a tree that would make it possible to determine—just from the *abstract mathematical structure* of this or that hypothetical fundamental theory of the world, just from (say) the *Hamiltonian* of this or that hypothetical fundamental theory of the world—whether or not the theory in question could actually accommodate the existence of tables and chairs and rocks and trees. But the prospects for an analysis like that, as far I can tell, do not look good.

Here's a very different approach: There are people who say that the right way of thinking about the connections between things like tables and chairs and rocks and trees (on the one hand) and things like particles and fields and marvelous points (on the other) is as *contingent* and *a posteriori* features of how nature *actually happens to work,* as something very closely akin

---

[8] Or (rather) the question of whether or not a fundamental theory like Theory 2 can accommodate particles and rocks and trees and haircuts and universities and toothbrushes, and so on, *given that Theory 1 can.*

to (that is) *causation,* or *production,* or *explanation,* or *natural law.* Call them (for reasons that will become apparent as we go on) the *libertines.*[9]

The libertines say that the question that we have been talking about here, the question at issue between the functionalists and the puritans, the question (that is) of whether there are tables and chairs and rocks and trees in Theory 2—is not, as it stands, *well posed.* They say that *neither* Theory 1 *nor* Theory 2—as they were spelled out at the beginning of this essay—entail *anything at all,* in and of themselves, about whether there are such things as rocks and trees and haircuts and so on. They say that there are substantive metaphysical facts about *what grounds what*—facts that are *over* and *above* and *apart* from anything that was mentioned in our earlier presentations of Theory 1 and Theory 2—that need to be appealed to in order to underwrite inferences from the fundamental to the nonfundamental. These additional facts bridge the gap (as it were) between the to-ing and fro-ing of the fundamental stuff and whatever facts there might happen to be about tables and chairs and haircuts in something like the way that the contingent dynamical *laws of nature* bridge the gap between (say) initial conditions and final ones.

Maybe this is worth belaboring a little further.

Both the puritans and the global functionalists agree that all that God had to do to create the entirety of the physical world was to create the fundamental physical stuff, and to write down

---

[9] It ought to be acknowledged here—just as it was acknowledged, in footnote 3, in the case of the puritans—that I am not altogether certain that any such people actually exist. I think that Jonathan Schaffer is a libertine in papers like "The Ground between the Gaps," *Philosophers' Imprint* 17, no. 11 (May 2017): 1–26. And maybe Ezra Rubenstein is too, in papers like "Grounded Shadows, Groundless Ghosts," *British Journal for the Philosophy of Science* 73, no. 3 (2022): 723–750. Anyway, it will be useful, for purposes of exposition, whether there actually turn out to be any of them or not, to have their doctrine on the table.

the fundamental physical laws, and to choose the initial conditions, and to sit back and let everything take its course. Both of them agree (that is) that the entirety of the physical facts are settled by the facts about what there fundamentally *is* and the facts about what that fundamental stuff *does*.[10] And both of them will agree that there are *a priori principles of reasoning* by means of which all of the physical facts—the facts about rocks and trees and tables and chairs and haircuts and so on—can be *read off* of the facts about what there fundamentally is, and about what that fundamental stuff does. The *disagreement* between puritans and the global functionalists has to do with what those a priori principles *say*, but not with their *existence*, or with their *necessity*, or with their *a priority*. The disagreement between the puritans and the global functionalists has to do (in particular) with what it actually *is*, or what it actually *takes*, to *be* a rock or a tree or a table or a chair.

The *libertines*, on the other hand, reject *both* of those positions. They think that there was a great deal *more* for God to do, in creating the physical world, than to create the fundamental stuff and the fundamental laws and the fundamental initial conditions. The libertines will agree with the puritans (mind you) that there is an explanatory and metaphysical *gap* between the motions of the marvelous point in Theory 2 and the existence of things like particles and tables and chairs and haircuts and lawsuits—but the libertines will also insist that there is just as *much* of a gap, and just the same *kind* of a gap, between the motions of the *particles* in Theory *1* and the existence of tables and chairs and haircuts and lawsuits. And these gaps are not susceptible of being bridged, so the libertines say, by means of

---

[10] And what I have in mind here, when I talk about "what that fundamental stuff does," is both what the fundamental stuff does and the *laws* of what the fundamental stuff does. For Humeans, of course, the latter will supervene on the former. For others it won't.

mere conceptual analysis, or by means of a priori principles of reasoning—they can only be bridged by appeal to additional and substantive and a posteriori facts, out there in the world, about what grounds what.

On this way of thinking, these "rules of grounding" are just one more category of things that need to be written down—like equations of motion, or laws of interaction, or spaces of possible states, or what have you—in order to specify a physical theory. So (for example) there are rules of this kind that could be added to Theory 2 in such a way that the resulting theory—call it Theory 2a—will accommodate particles and trees and rocks and tables and chairs and all of the macroscopic paraphernalia of our everyday experience of the world. And there are *other* rules of this kind that could be added to Theory 2 in such a way that the resulting theory—call it Theory 2b—will *not* accommodate particles or tables or chairs or golf balls, or (for that matter) *any* of the paraphernalia of our everyday macroscopic experience of the world. And there are still *other* rules of this kind that could be added to Theory *1* in such a way that the resulting theory—call it Theory 1a—gives us tables and chairs and golf balls and so on whenever and wherever there are particles moving table-wise and chair-wise and golf-ball-wise and so on, respectively (and *these,* of course, are just the rules that we usually and uncritically and unconsciously *take for granted* when we think about a Newtonian mechanics of particles). And there are rules of this kind that could be added to Theory 1 in such a way that the resulting theory—call it Theory 1b—notwithstanding that it has particles in it, and notwithstanding that those particles sometimes move around table-wise and chair-wise and golf-ball-wise—gives us *no* tables or chairs or golf balls or any of the paraphernalia of our everyday macroscopic experience. And there are rules of this kind that could be added to Theory 1 in such a way that the resulting theory—call it Theory 1c—gives us tables and chairs and golf balls and so on exactly *five feet to the left* of the particles moving table-wise and chair-wise and golf-ball-wise and so on, respectively.

The libertines say that Theories 1a and 1b and 1c present different and incompatible pictures of the world—in much the same way as (say) a version of Newtonian mechanics (fitted out with the familiar grounding rules of Theory 1a) in which the force of gravitation falls off as $1/r^2$, and a version of Newtonian mechanics (also fitted out with the grounding rules of Theory 1a) in which the force of gravitation falls off as $1/r^3$, present different and incompatible pictures of the world. And the same goes for Theories 2a and 2b. And the same goes, of course, for Theories 2a and 1a—even though *those* two theories turn out to be *empirically indistinguishable* from one another.

For the libertines (then) the question of whether Theory 2 can accommodate the three-dimensional paraphernalia of our everyday experience of the world (particles, and rocks, and trees, and so on) is not the locus of any deep philosophical perplexity—it's just a question of what particular *grounding rules* one *fits the theory out with.*

And so the libertines seem to be offering a straightforward and fundamental and metaphysically principled *way out* of the predicament of global functionalism—in a way that *scientific* functionalism (for example) does not. The libertines are certainly not going to be committed (for example) to the claim that if God were to place a single structureless point-like physical item at rest in a sufficiently high-dimensional fundamental space, and to stipulate that the single fundamental law of the evolution of the world is that *nothing ever happens,* she would thereby necessarily and unavoidably bring all of the tables and chairs and haircuts and universities of our everyday experience of the world into being. According to the libertines (remember) there is a great deal more that God needs to do—in creating a world—than merely creating the fundamental stuff and stipulating the dynamical laws and the initial conditions of that fundamental stuff. She also needs to stipulate the rules about *what grounds what*. And she is free to stipulate whatever such rules she likes—just as she is free (say) to stipulate whatever *dynamical laws* she likes—and until she does, *nothing whatever* is going to follow about whether or not there are tables or chairs or haircuts or universities.

But the libertines have new and distinctive troubles of their own, and those troubles can be traced back to precisely the freedom that I mentioned above—those troubles can be traced back (that is) to the doctrine that the connections between the fundamental and the nonfundamental are features of the way the world merely contingently happens to be.

The libertines say that Theories 1a and 1b and 1c (for example) represent meaningfully and intelligibly and substantially different ways that the world might contingently happen to be. But that seems to take it for granted that we have some kind of a grip on *what it is* to be a particle that applies across *all three* of those theories—that seems to take it for granted (that is) that we have some kind of a grip on what it is to be a particle that is relatively *independent* of what the relationship between particles (on the one hand) and tables and chairs and golf balls and so on (on the other) actually *is*. And that doesn't seem right. "Particle," after all, at least in the way we are using it here, is a technical term of physics—and the way in which that term is invariably *introduced* into our fundamental physical theories is *precisely* as a *tiny little piece* of a *table* or a *chair* or a *golf ball*, the way in which that term is invariably introduced into our fundamental physical theories (that is) is by means of locutions like: "Whenever a collection of these 'particles' are moving around in such-

and-such a way in such-and-such a spatial region, then there is a table or a chair or a golf ball in that same spatial region." And we have no grip at all on what it is to be a particle, as far I can see, that does not depend on that particular way of talking about them. And so it isn't clear how it could be the case—it isn't clear (that is) that there is even so much as a conceptual or metaphysical *possibility* of it's being the case—that particles are related to tables and chairs in anything like the ways that are envisioned in Theories 1b and 1c.[11]

And very much the same sorts of considerations would seem to apply to fundamental physical theories in general. Imagine that somebody—call her Darleen—says, "I have a new proposal for a fundamental physical theory of the world: The world con-

---

[11] A couple of parenthetical remarks are in order here:

First, it's easy to get mixed up, in the context of a discussion like the one we are having here, about what one means by terms like "metaphysical possibility" and "metaphysical necessity." The trouble (in a nutshell) is that the libertines think of their grounding rules as contingent features of the way the world happens to be—they think of them as something akin (in particular) to relations of *causation* or *production* or *natural law*—and they are consequently in the habit of referring to those rules as "metaphysical laws." But referring to those rules as "metaphysical laws" suggests that that they represent "metaphysical necessities," in much the same way that *physical* laws represent *physical* necessities—and that a *violation* of those rules would amount to a metaphysical *impossibility*, in much the same way that a violation of the laws of *physics* would amount to a *physical* impossibility—and all of this (in turn) can be hard to put together with the libertines' original insistence that the rules of grounding are contingent features of the way the world actually happens to be. Anyway, my business here is certainly not to sort any of that out—but merely to reassure the reader that *I myself* am using terms like "metaphysical necessity" and "metaphysical possibility"—both at this point in the text and throughout the whole of this essay—in the more familiar and more straightforward and more flatfooted manner of the vulgar. I am (that is) going to treat "metaphysical possibility" as encompassing everything that *makes sense*, everything that is susceptible of being coherently *entertained* or *imagined* or *conceived of*. And so (for example) the libertines will assert, and I will deny, that Theories 1b and 1c are—in *my* sense of "metaphysically possible"—metaphysically possible.

sists entirely of fundamental objects called *schmoozles,* whose physical states evolve in time in accord with such-and-such equations of motion. Tell me what you think!" And we tell Darleen that we have no idea what to think, because she has told us nothing whatever about how to *apply* her theory to the business of making predictions about the behaviors of the various macroscopic paraphernalia of our everyday experience of the world, and so we have no way to make a judgment about its empirical adequacy. And Darleen responds—apologetically—that she somehow neglected to mention, that she certainly *ought* to have mentioned, that whenever the schmoozles are behaving *this* way then we have a table, and that whenever the schmoozles are behaving *that* way then we have a chair, and so on. And now we understand one another. Now we have been presented with a *theory.* And now we are in a position to consider the question of how well or how poorly this theory succeeds as an account of our empirical experience of the world.

Second, to say that we have no *grip* on what it is to be a particle except by way of the thought that particles are little pieces of tables and chairs and golf balls is *not* to say that what it is to be a particle is to be a little piece—or even a *potential* little piece—of a table or a chair or a golf ball. There are plenty of things that we are accustomed to thinking of as particles, there are plenty of things that we are *right* to think of as particles—neutrinos, say, or photons, or gluons, or what have you—that are not the kinds of things that things like tables and chairs and golf balls can even *potentially* be made of. And what's going on in cases like that is simply an application of the *scientific* variety of functionalist analysis that we talked about in the previous section: We say (that is) that to be a particle is to be the sort of thing that *interacts* with *other* things—and in this case with other *particles*—in a certain particular sort of way, or in one of a certain particular *class* of ways. And we take it as antecedently given—at least for the sake of this particular conversation—that tiny little pieces of tables and chairs are cases of particles. And that provides the *anchor* for our local functionalist analysis. That snaps everything else into place. And so—although it is not any part of the *meaning* of "particle" that a particle is a tiny little piece of a table or a chair—it is nevertheless exceptionlessly and ineluctably *by way* of things like tables and chairs that we have whatever grip we do on what it is to be a particle.

The libertines will say that when Darleen was talking to us, at the end, about the schmoozles and the tables and the chairs, she was telling us about substantive and a posteriori "metaphysical laws" that connect schmoozles (on the one hand) to tables and chairs (on the other). But that feels (again) like a strange way to think about it. That suggests that we have some kind of a grip on what it is to be a schmoozle that is more or less *independent* of the relationship between schmoozles and tables and chairs. And that *surely* isn't right. We don't have a *clue*, after all, outside of the context of Darleen's proposal, what in the world a schmoozle might be. We have only just this minute *heard* of them. And it feels much more exact and much more correct and much more illuminating to say that when Darleen was talking, at the end, about the schmoozles and the tables and the chairs, she was telling us something about what *kinds* of things the schmoozles *are*. And this (as far as I can see) is the only kind of a grip that physics is ever in a position to *offer* us on the intrinsic character of the fundamental stuff of the world.[12]

Here's another way to make pretty much the same point: The libertines talk about the connection between the fundamental stuff of the world and the nonfundamental stuff of the world—as I mentioned above—as if it were something akin to *causation* or *production* or *explanation*. And that seems to misunderstand (yet again) the kind of grip we have on what it is to be a "particle" or a "marvelous point" or a "schmoozle." What becomes apparent—as soon as we reflect on the way in which terms like "particle" are actually introduced into our scientific picture of the world—is that there is simply not enough of a *conceptual distance* between "table" (on the one hand) and "schmoozles

---

[12] Aside, of course, from the purely *structural* and *mathematical* relationships between the various different kinds of fundamental stuff—the kinds of relationships (that is) that the global functionalists appeal to, the kind that are specified in Ramsey sentences.

behaving table-wise" (on the other) for the second to amount to anything like an illuminating *explanation* of the first.

Consider (for example) the following claims:

(1) A certain projectile has a certain position and velocity at a certain particular time t because it had a certain *other* position and velocity at some particular *earlier* time t'.
(2) There is a table in the room because there are particles behaving table-wise in the room.
(3) Opium puts you to sleep because it possesses a *virtus dormativa*.

(1), of course, is precisely the sort of claim that is paradigmatic of successful scientific explanation. And it seems essential to the kind of *understanding* that an explanation like this provides that the kind of conceptual grip we have on the position and the velocity of a projectile at any *one* time is neither *prior to* nor *parasitic on* the kind of conceptual grip we have on the position and the velocity of that projectile at any *other* time.

And (3) is the sort of claim that provides us with no genuine understanding at all. And it seems essential to the explanatory *failure* of a claim like (3) that possessing *virtus dormativa* is *analytically* related—or as close as anything ever actually *gets* to being analytically related—to putting you to sleep.

And the worry (in a nutshell) is that (2) seems closer to (3) than it does to (1).

And note (finally) that if all this is right—then the libertines are wrong to suggest that there can have been anything more for God to do, in creating the physical world, than to make the fundamental stuff, and to stipulate the laws of how that stuff *behaves,* and to pick out the *initial conditions* of that stuff. There simply isn't any *room* (once all of that is done) to specify the connections between the fundamental and the nonfundamental—because the only kind of grip that science is ever going to be in a position to offer us on the intrinsic characters of the fundamental constituents of the world is one that implies that those connections are *part and parcel* of those intrinsic characters. The only kind of grip that science is ever going to be in a position to offer us on the intrinsic characters of the fundamental constituents of the world (that is) is one that implies that those connections are matters of some kind of conceptual or metaphysical *necessity*.

Picking Up the Threads

Think back, with all this in mind, to the original presentations, at the very beginning of this essay, of Theory 1 and Theory 2. We are now in a position to see that there was an important disanalogy between those two presentations—one that might not have been apparent at the outset.

In the presentation of Theory 1, the fundamental objects were called "particles." And calling them "particles" is telling us something—something (that is) over and above the fact that they are point-like items that move around in a three-dimensional space in accord with the Hamiltonian in equation (1)—about the *kinds* of things that they *are*. Calling them "particles" is telling us something (in particular) about how they are connected with tables and chairs and people and golf balls and so on. It's telling us that the fact that the Hamiltonian in equation (1) allows these particles to move around table-wise and chair-wise and haircut-wise means that Theory 1 can accommodate the existence of *bona fide* tables and chairs and haircuts.

In the presentation of Theory 2, on the other hand, the fundamental object is referred to as a "marvelous point"—which tells us *nothing whatever* over and above the fact that it is a point-like item that moves around in a $3N$-dimensional space in accord with the Hamiltonian in equation (2)—and nothing (in particular) about how it is connected, or about whether it is connected *at all,* with the familiar paraphernalia of our macroscopic experience of the world. Calling it a "marvelous point" is exactly as informative as calling it (say) a "schmoozle."

And so—notwithstanding their superficial similarities of grammatical structure—our original presentation of Theory 2 turns out to have said a great deal *less* than our original presentation of Theory 1. And so the libertines are right to say that the question of whether or not there are rocks and trees and haircuts in the sort of world described by Theory 2 is not, as it stands, well posed. And they are right to say that there is a *gap* between the gyrations of the marvelous point in Theory 2 (on the one hand) and the affairs of tables and chairs and haircuts and so on (on the other). But they are *wrong* (as far I can see) to say that there is just as *much* of a gap, and the same *kind* of gap, between the motions of the particles in Theory *1* (on the one hand) and the affairs of tables and chairs and haircuts (on the other). And they are wrong to suggest that there needs to be anything more to physics than an account of what there fundamentally *is,* and what that fundamental stuff *does.* And they are wrong to suggest that there was anything more that God needed to do, in creating the world, than to *create* what there fundamentally is, and to *promulgate* the *laws* of what that fundamental stuff *does.* The trouble with Theory 2 is not that it lacks substantive metaphysical principles of *grounding*—the trouble with Theory 2 is (you might say) that it *names* what there fundamentally is without *saying* what there fundamentally is.

Suppose that somebody proposes to remedy this defect (then) by telling us *what kind of thing* this "marvelous point" is supposed to *be*—to wit: When the marvelous point moves around in its $3N$-dimensional space under the influence of a Hamiltonian like the one in equation (2), then there is a system of N

Newtonian particles, moving around in a three-dimensional space, in accord with a Hamiltonian like the one in equation (1).

Well—okay. One can *say* that. One can say (as it were) that this is the *intended interpretation* of the "marvelous point" in Theory 2. One can say that this is precisely the kind of thing that Theory 2 *stipulates* the marvelous point to *be*. But none of that does anything to settle the question of whether the existence of a marvelous point like that—call it a "*really* marvelous point"—amounts to a *genuine metaphysical possibility*.[13]

And *that* (I think) is the best and sharpest way of putting the question that this essay has—all along, and unbeknownst to itself—been about.

The puritans say that it is a matter of a priori conceptual analysis that the existence of a "really marvelous point" is not a genuine metaphysical possibility *at all*. And the global functionalists say that it is a matter of a priori conceptual analysis that there are *any number* of genuine metaphysical possibilities of the existence of a "really marvelous point"—they say (in fact) that it is a matter of a priori conceptual analysis that *any point-like item whatever*, no matter what intrinsic "kind" of point-like item it may happen to be, so long as it is moving around in its $3N$-dimensional space in accord with a Hamiltonian like the one in equation (2), *necessarily* amounts to a "really marvelous point." And the libertines (if I am understanding them correctly) will say that it is a matter of a priori conceptual analysis that any point-like item whatever, that is moving around in a

---

[13] This (if I understand her correctly) is precisely the question that Alyssa Ney is struggling with in her recent and excellent book *The World in the Wave Function: A Metaphysics for Quantum Physics* (Oxford: Oxford University Press, 2021), sec. 7.4. She calls it the problem of *constitution*—and the solution that she proposes seems to me to amount to a version of what I have been calling libertinism.

$3N$-dimensional space in accord with a Hamiltonian like the one in equation (2), no matter what intrinsic "kind" of point-like item it may happen to be, *can* amount to a "really marvelous point"—and that whether or not it *does* will depend on what particular *grounding relations* the world may contingently happen to have been fitted out with. And I have been doing my best to explain why all of them strike me as somehow wanting or confusing or wrong.

In Lieu of a Conclusion

Let me see if I can say, for whatever it may be worth, where this leaves me.

The question (again) is whether physics can even so much as coherently *entertain* the hypothesis that the tables and chairs and haircuts and universities of our everyday experience of the world emerge from the to-ings and fro-ings of fundamental stuff in the $3N$-dimensional space of Theory 2. The question (that is) is whether there is even so much as a conceptual or metaphysical possibility there *at all*. And what all of our foregoing tribulations seem to me to suggest is that this question is not likely to succumb to any *a priori* analysis, or to any distinctively *philosophical* kind of analysis—and that we might do better to approach it as a further question for physics *itself*.

Here's how I'm imagining that might work:

Suppose that we were willing to pretend, as we agreed to do at the outset, that there were solutions to the equations of motion associated with a Hamiltonian like the one in equation (1) in which there are particles moving table-wise and chair-wise and rock-wise and tree-wise and so on. In that case, the situation would be pretty cut and dried. Theory 1 would manifestly amount to the best possible fundamental physical theory of a world like ours. And there would be no good scientific reasons whatever for entertaining the hypothesis that the tables and chairs and rocks and trees of our everyday experience were actually realized by the to-ings and fro-ings of anything in a high-dimensional fundamental space like the one in Theory 2. And there would be no good scientific reasons whatever for supposing that the tables and chairs and rocks of our everyday experience of the world *could* be realized by the to-ings and fro-ings of anything in a high-dimensional fundamental space like the one in Theory 2—there would be no good scientific reasons (that is) for supposing that there was even so much as a *metaphysical possibility* or a *skeptical scenario* in which the tables and chairs and rocks of our everyday experience of the world were realized by the to-ings and fro-ings of anything in a high-dimensional fundamental space like the one in Theory 2. And there would—at the same time—be no good scientific reasons whatever for supposing the *contrary*.

So—if there *were* solutions of the classical equations of motion associated with the Hamiltonian in equation (1) in which collections of particles are moving table-wise and chair-wise and rock-wise and tree-wise and so on, and if we actually *lived* in the kind of world that those equations described, then the question of whether anything along the lines of Theory 2 could accommodate tables and chairs and rocks and trees would simply not be among the questions that we would have any way of productively *reasoning* about—it would simply not be among the questions that we would have any way of getting a *purchase* on.

But of course there aren't *really* any solutions to the classical equations of motion associated with the Hamiltonian in equation (1)—and (as a matter of fact) there aren't really any solutions to *any* classical equations of motion—in which collections of particles are moving table-wise and chair-wise and rock-wise and tree-wise and so on. And it turns out that the business of coming to grips with all *that*—which is (in a nutshell) the business of *quantum mechanics*—presents us with *any number* of ordinary scientific sorts of reasons for choosing a theory in which the tables and chairs and rocks and trees of our everyday experience of the world are associated with the to-ings and fro-ings of fundamental physical stuff in precisely the kind of high-dimensional space that we find in Theory 2.[14] And the upshot of the present considerations is just that there seem to be no *a priori* reasons, and that there seem to be no *conceptual* sorts of reasons, and that there seem to be no *distinctively philosophical* sorts of reasons, why those *ordinary scientific* sorts reasons ought to be overruled, or set aside.[15]

One more example: Suppose that somebody proposes that Theory 4 is the true fundamental theory of our world. I can't see that there are going to be any conceptual or a priori or distinctively philosophical reasons for rejecting a proposal like that. I think that the right thing to do, I think that right way

---

[14] The "fundamental physical stuff" in question here is, of course, the quantum-mechanical *wave-function*—and the "ordinary scientific sorts of reasons" I have in mind here are the ones I have described in "Elementary Quantum Metaphysics," in *Bohmian Mechanics and Quantum Theory: An Appraisal*, ed. J. T. Cushing, Arthur Fine, and Sheldon Goldstein (Dordrecht, Netherlands: Kluwer, 1996), 277–284; in "Primitive Ontology," in *After Physics* (Cambridge, MA: Harvard University Press, 2015), 144–160; and (especially and particularly) in Chapter 1 of this book ("A Guess at the Riddle").

[15] This is a conclusion that libertines will endorse—but for somewhat different reasons. The libertines (as I mentioned at the end of the previous section) will say that it is a matter of *a priori conceptual analysis* that *any point-like item whatever* that is moving around in a $3N$-dimensional space in accord with a Hamiltonian like the one in equation (2), no matter what intrinsic "kind" of point-like item it may happen to be, *can* amount to a "really marvelous point"—and that whether or not it *does* will depend on what particular *grounding relations* the world may contingently happen to have been fitted out with. And indeed, Ezra Rubenstein has argued, from the perspective of a libertine, for precisely this conclusion in "Grounded Shadows, Groundless Ghosts."

to be *skeptical* about a proposal like that, is to wonder aloud whether Theory 4 can really be the *best* and most *perspicuous* and most *explanatory* scientific account we have, or will eventually discover, of the behaviors of tables and chairs and rocks and trees.[16]

Glossary of Theories

Theory 1: A classical theory of N fundamental particles moving around in a fundamental three-dimensional Euclidian space under the influence of the Hamiltonian in equation (1).

Theory 2: A classical theory of a single fundamental point-like physical item—called the "marvelous point"—moving around in a fundamental $3N$-dimensional Euclidian space under the influence of the Hamiltonian in equation (2).

Theory 3: A classical theory of a single fundamental point-like physical item floating around in a $6N$-dimensional fundamental physical space.

Theory 4: A classical theory of a single structureless point-like fundamental physical item that remains permanently at rest in a fundamental Q-dimensional space.

Theory 5: A classical theory of the motions of $Q/6$ fundamental Newtonian particles floating around in an infinite three-dimensional Euclidian space and having no interaction with one another *at all*.

Theory 6: A classical theory of the motions of $Q/6$ fundamental Newtonian particles floating around in a fundamental three-dimensional space that interact with one another only by way of very short-range mutually repulsive forces—like billiard balls.

---

[16] I owe this way of putting it to a conversation with Sidney Felder.

Theory 4*: A classical theory of a single point-like fundamental physical item in a $(Q+1)$-dimensional space that always moves in the *f*-direction with velocity 1.

Theory 2a: Theory 2 fitted out with contingent *grounding rules* that allow it to accommodate particles and trees and rocks and tables and chairs and all of the macroscopic paraphernalia of our everyday experience of the world.

Theory 2b: Theory 2 fitted out with *other* contingent grounding rules that do *not* allow it to accommodate particles or tables or chairs or golf balls, or (for that matter) *any* of the paraphernalia of our everyday macroscopic experience of the world.

Theory 1a: Theory 1 fitted out with contingent grounding rules that give us tables and chairs and golf balls and so on whenever and wherever there are particles moving table-wise and chair-wise and golf-ball-wise and so on, respectively (and *these*, of course, are just the rules that we usually and uncritically and unconsciously *take for granted* when we think about a Newtonian mechanics of particles).

Theory 1b: Theory 1 fitted out with contingent grounding rules such that—notwithstanding that this theory has particles in it, and notwithstanding that those particles sometimes move around table-wise and chair-wise and golf-ball-wise—gives us *no* tables or chairs or golf balls or any of the paraphernalia of our everyday macroscopic experience.

Theory 1c: Theory 1 fitted out with contingent grounding rules that give us tables and chairs and golf balls and so on exactly *five feet to the left* of the particles moving table-wise and chair-wise and golf-ball-wise and so on, respectively.

The "amended" version of Theory 2: A version of Theory 2 in which we are explicitly told that the "marvelous point" is to be understood as a "really marvelous point." A version of Theory 2 (that is) in which we are explicitly told that when the marvelous point moves around in its $3N$-dimensional space under the influence of a Hamiltonian like the one in equation (2), then there is a system of N Newtonian particles, moving around in a three-dimensional space, under the influence of a Hamiltonian like the one in equation (1). The question at the heart of this essay is whether or not the existence of a marvelous point *like that*—a "really marvelous point"—is a genuine metaphysical possibility.

ACKNOWLEDGMENTS

There are many people to thank.

I am so enormously grateful—to begin with—for having had the luxury of playing with these ideas, over the course of many years, in front of the brilliant and creative and generous audiences at graduate seminars I have taught in the philosophy department at Rutgers University. These seminars were often presented in collaboration with my longtime friend, and my great teacher, Barry Loewer—and they were often attended by the likes of Ted Sider and Jill North and Ezra Rubenstein and Veronica Gomez and Isaac Wilhelm and Chloe Park and Eddy Chen and Jonathan Schaffer and Nikitas Koutoupes and Saman Majid and Zee Perry and Ye-Eun Jeong and any number of other wonderfully patient and curious and talkative and insightful people—and all of them nurtured and criticized and worried over and improved these ideas in ways that I cannot begin to enumerate.

And I have had many important and encouraging conversations about this stuff with Sidney Felder, and with Trevor Teitel, and with Jenann Ismael, and with my brother Marc Albert. Jenann has herself been thinking, for a long time now, about ideas that are more or less along these lines—and all four of them have been willing to wade through great chunks of this material, in various stages of its interminable gestation, and have done much to make it better.

And I am grateful, as I have always been, to Tim Maudlin—for his fierce and loyal and frightening and untiring and invaluable critique.

And I am grateful to Ian Malcom and Sharmila Sen and Rachel Field and Aaron Wistar and Tim Jones and Eric Mulder of Harvard University Press.

And to countless others too. Thank you all.

INDEX

Albert, D., x, 7, 44n4, 82n8
Albert, M., 127
analytical mechanics, 12n1, 98

Barrett, J., x
Bell, J., ix, 3
Bell's inequality, 30
Bohm, D., 3
Bohmian mechanics, ix, 5, 7, 28, 35, 44, 50–53, 55, 73–76, 80–85, 122
Bohr, N., x, 1–8, 50
Boltzmann, L., 4

causation, 108, 113n11, 115
collapse of the wave-function, ix, 30; effective (in Bohmian mechanics), 36; in the GRW theory, 78
configuration space, 17–18, 52–53n8, 89–90
Copenhagen interpretation of quantum mechanics, 1–5

Darwin, C., 4
determinism, 66–69, 72, 76–78, 80–83
double-slit experiments, 7n3, 49

Einstein, A., ix, 3, 4
electromagnetism, 6, 9, 24, 41, 59, 69, 71–72, 74–76, 79–80, 84
energy, 11, 12, 13, 15, 32, 46, 63n1
Everett, H., III, 3
Everett interpretation of quantum mechanics, 6, 7, 84, 85, 91n4, 101n5. *See also* many-worlds interpretation of quantum mechanics

Felder, S., 123n16
Feynman, R., 48n7, 72n5
Field, H., 65n2
fields, 6, 10, 16; in configuration space, 36–41; elimination from the fundamental ontology, 71–80

INDEX

Goldstein, S., 89, 91n3
GRW theory of the collapse of the wave-function, 5–7, 44n5, 50, 57, 78, 84, 85n10

Hamiltonian, 11–14, 24, 27, 32, 34, 36–40, 46, 48, 55–58, 89, 94, 103, 107, 117, 119–125
heat, 91–92, 103, 107
Heisenberg picture (of quantum mechanics), 98
Hilbert space, 46, 98

linear algebras, 46
linear dynamical laws, 35, 39, 52, 55
locality and nonlocality, ix, 48, 62, 63

many-worlds interpretation of quantum mechanics, ix, 34, 45, 50, 57, 91. *See also* Everett interpretation of quantum mechanics
Maxwell, J. C., 4, 6, 9, 24, 59, 71, 72, 74, 75, 76, 79, 80, 84
Maudlin, T., x, 91n4
measurement, 34–35, 40–46, 50–51, 56, 79, 84
measurement problem, ix, 1–7, 35, 44
Mentaculus, 76

Ney, A., x, 119n11

observables, 40–46, 98

past hypothesis, 81, 82n8
phase space, 93–95, 98, 100, 104
probability distributions, 76–83

relativity, 9, 59, 68

Schrodinger, E., ix, 3
Schrodinger equation, 81, 85
separability, 47
significables, 44n4
space of elementary physical determinables (definition of), 22–23
space of ordinary material bodies (definition of), 17
state vector, 39
statistical mechanics, 76–84
superposition, ix, 36, 48
supervenience, 21, 23, 26

theories: Theory 1, 88; Theory 1a, 110; Theory 1b, 110; Theory 1c, 111; Theory 2, 89; Theory 3, 94; Theory 4, 98; Theory 4*, 102; Theory 5, 99; Theory 6, 99; Amended Version of Theory, 2, 118–119

Von Neumann, J., ix

Wallace, A. R., 4
Wallace, D., 100n5
wave-function, ix, 6, 8, 9, 10, 29, 35, 40, 46, 49–57, 73–75, 78, 80–87, 91n4, 122n14
Wheeler, J., 72n5
Wigner, E., ix